Physics is ...

The Physicist explores attributes of physics

Physics is ...

The Physicist explores attributes of physics

F Todd Baker

Professor Emeritus, Department of Physics and Astronomy,
University of Georgia, Athens, GA, USA
and
The Physicist at www.AskThePhysicist.com

Morgan & Claypool Publishers

Rights & Permissions
To obtain permission to re-use copyrighted material from Morgan & Claypool Publishers, please contact info@morganclaypool.com.

ISBN   978-1-6817-4445-2 (ebook)
ISBN   978-1-6817-4444-5 (print)
ISBN   978-1-6817-4447-6 (mobi)

DOI   10.1088/978-1-6817-4445-2

Version: 20161201

IOP Concise Physics
ISSN 2053-2571 (online)
ISSN 2054-7307 (print)

A Morgan & Claypool publication as part of IOP Concise Physics
Published by Morgan & Claypool Publishers, 40 Oak Drive, San Rafael, CA, 94903 USA

IOP Publishing, Temple Circus, Temple Way, Bristol BS1 6HG, UK

For my many mentors and colleagues who have, over more than 50 years, inspired in me an unquenchable love of physics.

Joe Priest, Dave Griffing, George Arfken, Bob Tickle, Ted Hecht, Helmut Baer, Charlie Glashausser, Ted Kruse, Gary Love, and many others

Contents

Appendices

Preface

I am *The Physicist!* Since 2006 I have run a website, www.AskThePhysicist.com, where I answer questions about physics. The site is not intended for answering highly technical questions; rather the purpose is to answer, with as little mathematics and formalism as possible, questions from intelligent and curious laypersons. For several years before my retirement from the University of Georgia I ran a similar Q&A site for the Department of Physics and Astronomy there. Over the last dozen years I have answered nearly 6000 questions online and uncounted more by brief email replies. I have found this very rewarding because it is an extension of my more than 40 years' experience teaching and because I learn something new almost every day. The questions I receive reveal what aspects of physics interest people and what principles they do not grasp. They reveal a widespread thirst to understand how physics describes, on many levels, how our universe works. It is gratifying that the site has on the order of 50–100 000 visits per month, far more than the number of questions asked; I interpret this to mean that there are many visits by people who simply like to read and learn.

This is the third book of the *Ask The Physicist* series. The first book in the series, *From Newton to Einstein: Ask the physicist about mechanics and relativity*, focused on classical mechanics and relativity. The second book, *Atoms and Photons and Quanta, Oh My! Ask the physicist about atomic, nuclear, and quantum physics*, focused on topics in modern physics. The first two books were written such that the structure was mainly in chapters covering areas of physics—Newtonian mechanics, special relativity, electromagnetism and light, quantum physics, nuclear physics, etc. Furthermore, there was an objective to explicate the topics, to explain and 'teach' as well as to illustrate the topics with selected Q&As. When embarking on this third book, I felt that well was dry—a new structure was needed. I decided to structure the book around chapters exploring the attributes of physics—practical, beautiful, surprising, cool, frivolous. The first two books have chapters categorized by nouns, but this book has chapters categorized by adjectives. Also, since I did my 'teaching' in the first two books, there is less explication in the third book; so, if you are not familiar with the fundamentals of physics (say, high school level physics), you might find this book less accessible than the first two. To address this, appendices compiled from parts of *From Newton to Einstein* have been included in this book; appendix E is the introduction to Newtonian mechanics from chapter 1 of that book. If you have read the first two books, or even one of them, you should be well prepared for this one. And it is structured in such a way that there is nothing to prevent you from skipping over material that you find a little hard or uninteresting.

For those of you familiar with the first two books, be assured that a chapter with 'silly stuff' is also included in the third volume. To those new to the series, I like to include a chapter with a few examples of questions from people who think I am a psychic and from crackpots who think they have the most brilliant theory ever of the universe, just for a little levity.

This book is designed for a layperson, and is mostly heavy on 'concepts' rather than 'formalism', but I strive to keep the physics correct and not 'watered down'. Learn, enjoy, and have fun!

Acknowledgements

With the internet at my fingertips and a few classic texts I have held on to, I can usually find the answer to any reasonable layman's question in a few minutes— pretty much a one-man band. I am particularly indebted to the ROMEOs (retired old men eating out) with whom I frequently discuss interesting *Ask the Physicist* questions at our Wednesday lunches—Gary Love, Bob Anderson, Bob Wood, Alan Edwards, and Richard Meltzer. I should also acknowledge the visitors to the site who have submitted scores of questions which have challenged me to seriously research the topic or brush up on some physics I have not thought about for years; I learn something new or relearn something old almost every day.

Author biography

F Todd Baker

The Physicist is F Todd Baker. He received AB and MA degrees from Miami University and a PhD degree from the University of Michigan. His area of research is nuclear physics and he has published more than 70 articles in refereed journals as well as made numerous presentations at conferences and workshops. He has more than 35 years of college and university teaching experience. In 2006 he retired from the University of Georgia where he taught and performed nuclear physics research for 32 years. Previously he held a postdoctoral research associate position at Rutgers University and teaching positions at Carroll College (Wisconsin) and St Lawrence University. He now lives in Athens, Georgia with his wife Sara in a 106-year-old house mainly restored by him and decorated and landscaped by her. He has four beloved children aged 20–47 years. He enjoys bicycling around town, playing violin, cooking and baking, outdoor activities, DIY projects, film, music of many genres, working puzzles, reading mainly European murder mysteries, and hanging out in coffee houses. His curriculum vitae may be seen at www.ftoddbaker.com/cv.html.

Physics is ...
The Physicist explores attributes of physics
F Todd Baker

Chapter 1

Physics is practical

1.1 Introduction

This first chapter seeks to demonstrate that physics is practical, useful in everyday life, and illustrative of why and how so many things around us become comprehensible in terms of the laws of physics. Read on and enjoy!

1.2 Physics in sports

Many books with titles like *The Physics of Baseball*, *The Physics of Tennis*, *The Physics of Gymnastics*, and *The Physics of Ice Skating* have been published. A great deal about the subtleties and properties of athletic endeavors can be readily understood using the laws of physics.

The 2015 American Football Conference Championship Game was rocked by a scandal, colloquially known as 'deflategate', involving the inflation pressures of the footballs used in the game. It was alleged that the balls used by one team were tampered with by personnel from the other team. The question hinged on whether there is significant change in pressure if a ball inflated at room temperature is then moved to a much cooler temperature where the game is played. Physics! The following question was about this incident.

> **Question:** In light of the recent deflated football scandal, is there a way to mathematically calculate the change in pressure as the temperature inside the ball changes? Would you assume the volume of the bladder inside the ball to change very little? (see figure 1.1)
>
> **Answer:** Yes, the pressure change can be calculated using the ideal gas law, $PV/(NT) = $ constant; here P is pressure (*not* gauge pressure), V is the volume, N is the amount of gas, and T is the *absolute* temperature. Assuming that V and N remain constant, $P/T = $ constant. Let's do an example. Suppose that the ball is filled to a gauge pressure of 13 psi when the temperature is 70 °F. The absolute

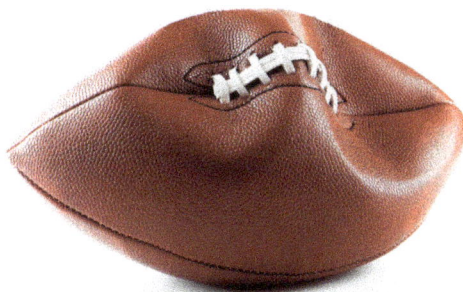

Figure 1.1. A really deflated football. Copyright: Africa Studio/shutterstock.

pressure is 13 psi plus the atmospheric pressure of 14.7 psi: $P_1 = 13 + 14.7 = 27.7$ psi. The temperature in kelvins (absolute) is 70 °F = 294 K. Now suppose we cool the football to 10 °F = 261 K. Then, $27.7/294 = P_2/261$, $P_2 = 24.6$ psi, and the resulting gauge pressure is $24.6 - 14.7 = 9.9$ psi. It is clearly important to fill the ball at the temperature at which it will be played.

Added note: An article in the 30 January 2015 *New York Times* came to essentially the same conclusion as I did here. My answer was posted on 26 January 2015. I was astounded to read in that article, though, that initial calculations *by physicists* had applied the ideal gas law using gauge pressure rather than absolute pressure! Shame on them!

I am proud of having beaten the *New York Times* to the first published *correct* analysis of this problem!

Another popular sport in which physics plays an important role is baseball. One of the reasons why physics is important is that the speeds at which baseballs typically travel are large enough for the effects of air drag to become important. The following question is an example of the importance of forces encountered because of the ball's interaction with the air.

Question: I am an avid sports fan and I have often wondered if the experts may be wrong about the myth of the rising fastball in the game of baseball. I played baseball for over 20 years and I can tell you that the ball does appear to rise when certain pitchers throw hard and put a heavy backspin on the ball. I have been told that experts say it is nothing more than a visual trick your eyes play on you because a rising fastball is considered to be physically impossible. I can tell you first hand that a softball pitcher I know can throw a ball that rises after being thrown on a straight trajectory. I suspect the Magnus effect may have something to do with the anti-gravitational behavior of the ball. Do you think this could be what causes the ball to appear to rise as it travels, or is it just our perception?

Answer: There is such a thing as a rising fastball, but it does not actually rise; it simply falls more slowly than a nonspinning ball does. An experienced hitter knows intuitively what a normal fastball does and, when presented with a rising fastball, he will swear that it rose because it actually fell less. Incidentally, by rise

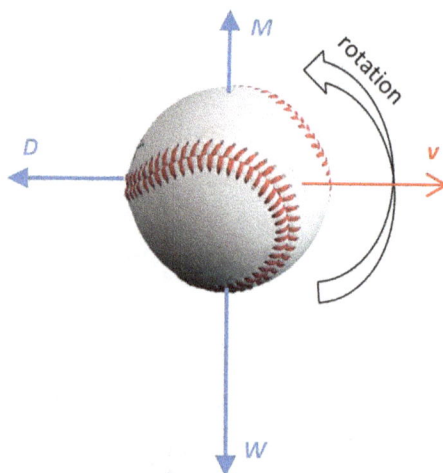

Figure 1.2. A 'rising fastball'.

or fall, I am talking about the direction of the acceleration. So a ball which is thrown at an angle above the horizontal is obviously rising, but it rises at a decreasing rate of rise until it reaches the peak of its trajectory and then begins back down; the rising fastball will actually rise farther. Then why do we say it is a myth? It is easiest to understand by looking at a ball thrown purely horizontally. Can spin cause the ball to actually go upwards? To answer this, you need to think about all the forces on a pitched ball, as shown in figure 1.2. There is the weight, *W*, which points vertically down and causes the ball to accelerate downward (a horizontally thrown 90 mph fastball falls about four feet on the way to the plate); there is the drag, *D*, which points opposite the direction of flight (the velocity, *v*, indicated in figure 1.2) and tends to slow the ball down (a 90 mph fastball loses about 10 mph on the way to the plate); and there is the Magnus force, *M*, which, for a ball with backspin about a horizontal axis points perpendicular to the velocity and upward. If the ball is moving horizontally the only way it could rise is for the Magnus force to be larger than the weight. Measurements have been done in wind tunnels and it has been found that if the rotation is 1800 rpm, about the most a pitcher could possibly put on it, the ball would have to be going over 130 mph for the Magnus force to be equal to the weight. When you say 'thrown on a straight trajectory', you cannot mean it left the hand horizontally, because it would hit the ground before it got to the plate; for a fast pitch like that it is impossible to accurately judge the initial angle of the trajectory.

A little more detail about the mechanism for the Magnus force is given in the next question.

Question: Consider a baseball pitched with a spin around the vertical axis. To be precise, let the ball's initial direction be southward and let the direction of its spin be clockwise when observed from above. Because of aerodynamic effects, the spinning ball will…

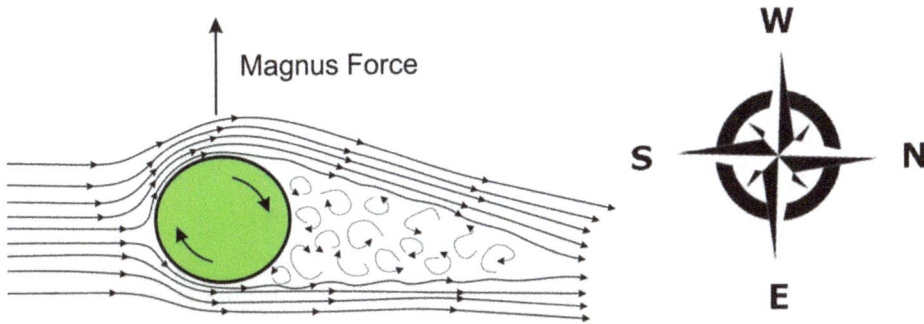

Figure 1.3. Magnus effect. Rdurkacz/CC-BY-SA-3.0 (https://creativecommons.org/licenses/by-sa/3.0/deed.en)

Answer: ...be deflected west. The general idea is shown in figure 1.3. The wake is deflected east in the figure for the ball moving to the south. The reason for the force that the ball experiences is Newton's third law. If the air is being deflected east, the ball must be exerting a force on it; if the ball exerts a force on the air, the air must exert an equal and opposite force on the ball. This force is called the Magnus force. There is also a Bernoulli force on the ball, also west, because the air goes faster over the top than the bottom and therefore there is lower pressure at the top; I believe that this is much less important than the Magnus force (although most elementary physics texts say that the Bernoulli force is the reason for the curve ball). Examples of pitches which break laterally are the slider and the screwball.

Another sport where the effects of the air are important is golf. You probably know that the reason for dimples on a golf ball is to, surprisingly, reduce air drag so the ball can go farther.

Question: Why does a golf ball have dimples?
Answer: How can we minimize the air friction for a ball? Our first inclination is to say it should be as smooth as possible so that it will slip smoothly through the air. But, counterintuitive though this may be, this expectation breaks down, particularly at high speeds, and it is advantageous to induce turbulence. To illustrate how turbulence affects drag and how smooth is sometimes not good, consider the golf ball. As you have noted, a golf ball has dimples. The purpose of these dimples is to *reduce* air drag. As shown on the left in figure 1.4, a smooth ball at a high velocity has a large turbulent volume behind it; because the pressure in this turbulent volume is significantly lower than the pressure on the front of the ball, this contributes to there being a large net drag force. If golf balls were nice and smooth, they would die and fall very much sooner than a dimpled golf ball does. The ball on the right shows the effect of the dimples; the rough surface induces a layer of turbulence that actually makes the ball 'slipperier', which causes the flow around the ball to come back together and reduces the volume of turbulence

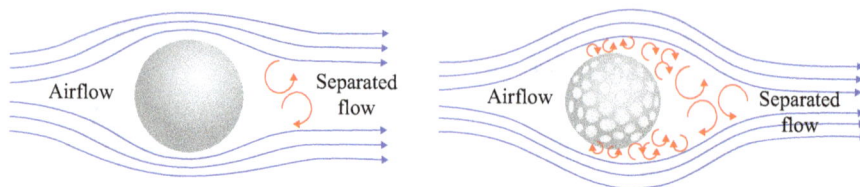

Figure 1.4. A golf ball with and without dimples.

contributing to drag. (The right side of figure 1.4, with most of the turbulence erased, shows what the smooth-ball air flow would look like at low speeds.) The hairs on a tennis ball serve the same purpose. One other example is the net you sometimes see replacing the tailgate of a pickup truck to reduce the drag the tailgate causes. This, it turns out, is a complete fraud. With the tailgate closed a bubble of still air forms in the bed of the truck, which deflects the air smoothly over the rear of the truck.

A golf ball can also spin. So one desirable thing to do is to hit the ball a little low so that it acquires backspin like the rising fastball above and therefore flies farther. But, if you hit the ball a little high instead, the ball will have forward topspin; the Magnus force will be down and the ball will dive down rather dramatically, resulting in a very short shot. It is also possible to impart spin about a vertical axis so that the Magnus force is horizontal, which causes the ball to be deflected left or right depending on the direction of the rotation; this is discussed in the next question.

Question: In golf, if I hit a ball very hard and then I hit one very softly, is the one hit very softly more likely to move or sway in its straight path?
Answer: You refer to 'its straight path'. No golf ball goes in a straight path, so I assume you mean that it does not curve left or right; such a ball, if not curving, would have a projected path on the ground (like the path of its shadow) that is straight. For a right-handed golfer, a ball which curves right is called a *slice* and one which curves to the left is called a *hook*; these have opposite spins. Neglecting the possibility of wind, the reason that a ball curves is because it has spin. But now it gets complicated because:
1. the hard-hit ball is in the air much longer than the softly hit ball;
2. the lateral force causing the curve depends on both the rate of spin and the speed of the ball, so the hard-hit ball will experience more lateral force than the softly hit one if they have the same spin;
3. even if the slow ball has a bigger lateral force, the fast ball is likely to be deflected a greater distance because of its longer flight time;
4. a lateral wind will exert the same force on both, but the fast ball will be deflected farther because of the longer time.

So, you see, there is no simple answer. To avoid curving, learn to hit the ball without imparting significant spin!

If you examine the surface of a basketball, you will see that it has little bumps on it. Is this for the same reason that the golf ball has dimples or the tennis ball has fuzz? The following question addresses this.

Question: Why is there a slightly rough surface on a basketball? Does this affect the static friction acting between your hand and the ball during a shot?
Answer: The dimples on a golf ball and the fuzz on a tennis ball reduce air drag and thereby allow the ball to go faster and farther. However, the speeds of these balls are much larger than the speeds encountered by basketballs and this cannot be the reason for the bumps on a basketball. A little research reveals that the purpose is just to make the ball easier to grip and handle, as you speculated.

While we are on the topic of golf balls, the following question is one of my all-time favorites. It involves not so much the game of golf, but rather an examination of a product designed to improve your game; I found the product to be useless.

Question: I received a novelty gift that purports to find the 'balance point' of a golf ball by spinning it up to 10 000 rpm. After 10–20 s the ball reaches an 'equilibrium' spin and a horizontal line is marked that indicates the 'balance axis'. The assumption is that on tees and greens you orient the ball to put the line vertically along the intended path so the center of gravity (CoG) is rolling/spinning over the target line and thus minimizing the potential effects of the CoG being on the side and potentially causing a 'wobble'. Putting the ball at different starting orientations in the device doesn't matter. It does tend to find the same equilibrium spin after a time, so it is consistent. I decided to mark a dozen balls using the device and then put them in a container of salt water to compare it to finding the CoG using a buoyancy test. Put in enough salt and eventually the golf balls will float and reorient to put the CoG at the lowest possible position in the solution. Out of 12 balls, only one had the previously marked axis running through the top of the ball. The others were all off by somewhere in the 40–45 degree range. The physics of the buoyancy test seem pretty straightforward and understandable to me. What is happening in the spin device is less clear. Can you explain the difference in the two tests? Is there a different 'balance point'/CoG that is being located by the spin device? And to preempt your first obvious statement, yes, I understand that none of this has a significant impact on my golf game compared to all of the other variables at play.
Answer: First, let's consider the physics of the spinning method. The claim of the manufacturer, Check-Go Pro, is that the CoG of the ball will seek a horizontal plane for high rotational speeds because of the centrifugal force; then, assuming that the CoG is in this plane, you use a marker to mark this 'equator' as the ball is spinning. Then, when you putt, aligning the equator vertically and pointing at the

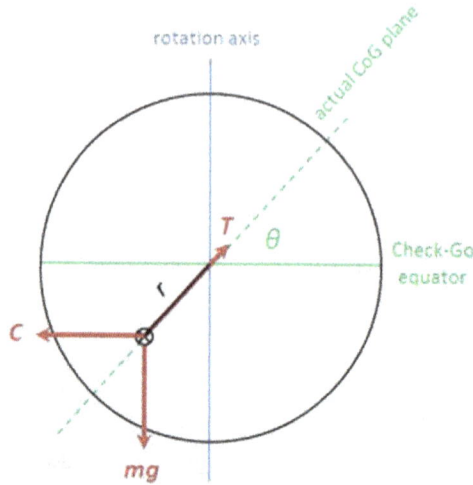

Figure 1.5. Forces on a golf ball rotating about a vertical axis. The location of the CoG is at a distance r from the center.

hole, the ball will roll true. But, will the CoG end up in a horizontal plane passing through the center of the ball? What I show here is that the answer is no. Refer to figure 1.5. If the CoG is a distance r from the geometrical center of the ball, has a mass m, and is spinning with an angular velocity of ω, the CoG experiences (in the rotating frame) three forces: the weight mg, the force T holding it in place, and the centrifugal force $C = mr\omega^2$. There will be an angle θ where the sphere is in equilibrium. Summing the torques about the center of the sphere, $mgr\cos\theta - mr^2\omega^2\sin\theta = 0$, so $\tan\theta = g/(r\omega^2)$. In words, the CoG is not in the horizontal plane ($\theta = 0$) unless $\omega = \infty$. (For physics aficionados, this is just the spherical pendulum problem.)

Now, the questioner found that for most balls, $\theta \approx 45°$, so $\tan\theta \approx 1$. Taking $g \approx 10$ m s^{-1} and $\omega^2 = (10\ 000\ \text{rpm})^2 \approx 10^6$ s^{-2}, I find that $r \approx 10^{-5}$ m. This is very small, 1/100 mm, but demonstrates that the spinner will not find the correct plane for very small r. Technically, it never finds the correct plane, but, as can be seen in figure 1.6, for $r > 0.2$ mm, $\theta < 2°$. Since it is my understanding that modern golf balls are very homogenous, this device will not be useful for nearly all balls.

Next I should address the question of whether such a small r will be detectable by the floating method. We can use figure 1.5, except with $C = 0$. The mass of the ball is about 0.046 kg and the radius of the ball is about $R = 0.021$ m. Taking $r = 10^{-5}$ m, the torque about the center of the ball is $\tau = mgr\cos\theta = 4.6 \times 10^{-6}\cos\theta$ N · m. The moment of inertia of the sphere is $I = 2mR^2/5 = 4.1 \times 10^{-5}$ kg · m^2. So, the angular acceleration is $\alpha = \tau/I = 0.11\cos\theta$ s^{-2}. This means that, if you start at $\theta = 0°$, after 1 s the angular velocity will be about $0.11 \times 180°/\pi \approx 6°$ s^{-1}, easily detectable I should think.

The bottom line here is that the questioner discovered that only one of the 12 balls was not, for all intents and purposes, perfectly 'balanced'.

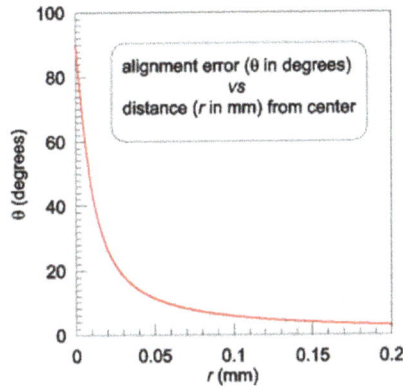

Figure 1.6. Alignment error vs distance from center.

All this is truly academic, though, since I am sure nobody really thinks that a ball with its CoG less than 0.2 mm from the center of the sphere will behave in any measureable way differently from a perfect ball. So the Check-Go Pro does no harm to your game, it just does no good unless you happen to have a really off-center CoG. If you do both measurements, you can locate surprisingly precisely where the CoG is, with θ giving you r and the vertical giving you the direction of the line between the CoG and the center of the sphere.

Knowing the equations of kinematics for uniform acceleration allows you to predict the trajectory of a ball easily, if you neglect the effects of air. This is a reasonable approximation if the ball is fairly massive, the time of flight small, and the speed not too large. Those are the assumptions I made in the answer to the following question about a soccer-ball free kick.

Question: I am currently doing a research paper on the perfect free kick. Could you find an equation that suits the following variables? The soccer ball is kicked from the origin of a coordinate system with an unknown velocity such that it passes through the points $(x, y) = (9.15$ m, 2.25 m$)$ and $(x, y) = (22.3$ m, 2.22 m$)$. How can I find the magnitude and direction of the initial velocity? Just having an equation to help me work with would be very nice.
Answer: The equations of motion for a projectile which has an initial velocity with magnitude v_0 and an angle relative to the horizontal θ are $x = v_{0x}t$ and $y = v_{0y} t - \frac{1}{2}gt^2$, where $v_{0x} = v_0 \cos\theta$, $v_{0y} = v_0 \sin\theta$, t is the time, and $g = 9.8$ m s^{-2}. Solving the x-equation for t, $t = x/v_{0x}$, and putting t into the y-equation, $y = (v_{0y}/v_{0x})x - \frac{1}{2}g(x/v_{0x})^2$. Since you have two (x, y) data points, you have two equations with two unknowns, (v_{0x}, v_{0y}). The algebra is tedious, but the result is that $v_{0x} = 21.0$ m s^{-1} and $v_{0y} = 7.30$ m s^{-1}; $v_0 = 22.2$ m s^{-1}, $\theta = 19.2°$. To check my answer I drew the graph shown in figure 1.7 (note the different x and y scales); it looks like my solution passes pretty close to the data points.

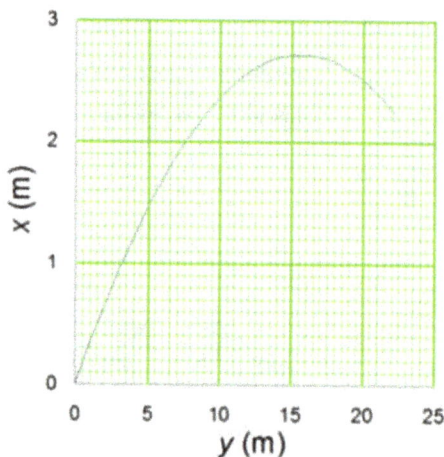

Figure 1.7. A 'perfect' soccer free kick.

At the time I answered this question, I simply neglected air drag. But, was that really a good approximation? I should have checked! The force of air drag at the time the ball is kicked can be roughly approximated as $F = \frac{1}{4}Av^2$, where A is the cross-sectional area and $v = 22.2$ m s^{-1} (this approximation only works for SI units). The radius of a soccer ball is about 0.1 m, so $A \approx 0.031$ m^2. Then $F \approx 3.9$ N, and the mass of a soccer ball is about $m \approx 0.45$ kg, so the acceleration, using Newton's second law, is $a = F/m = 8.6$ m s^{-2}. This is far from negligible! The air drag force is almost as large as the weight of a soccer ball, $W = mg = 4.4$ N. So, using my answer to make the kick would probably result in the ball not even making it to the goal! A much stronger kick would be needed. There is a lesson here—when making approximations, always check that they are reasonable. *Mea culpa!*

Another sport about which I often receive questions is cycling. The following question addresses the reason for a common bike accident.

Question: When the brakes are suddenly applied to a fast-moving bike, the back wheel leaves the ground. Why?

Answer: The 'free-body diagram' showing the pertinent forces if the back wheel has not left the ground is shown in figure 1.8. The weight mg acts at the center of gravity of the bike + rider and each wheel has a normal force (N) and frictional force (f) from the ground. The bike has an acceleration a in the direction of the frictional forces, so $f_1 + f_2 = ma$. The system is in equilibrium in the vertical direction, so $N_1 + N_2 - mg = 0$. The bike is also in rotational equilibrium, so all the torques about any axis must be zero; summing torques about the front axle, $Rf_1 + Rf_2 + DN_2 - dmg = 0$, where R is the radius of the wheel, D is the distance between the axles, and d is the horizontal distance between the front axle and the center of gravity. Now, suppose that the rear wheel is just about to leave the ground; then $N_2 = f_2 = 0$. The three equations then become $N_1 - mg = 0$, $f_1 = ma$, and $Rf_1 - dmg = 0$. Putting the second equation into the third and solving for the acceleration, $a = g(d/R)$; if you are slowing down any faster than this, your rear

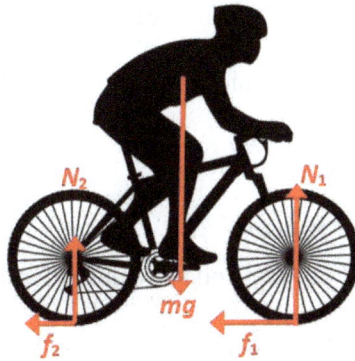

Figure 1.8. A cyclist.

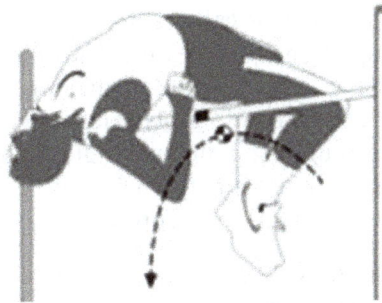

Figure 1.9. The Fosbury flop.

wheel will lift off the ground and the bike will no longer be in rotational equilibrium. If a is really big, you will keep rotating until your center of gravity is forward of the front axle; then you will not be able to stop the bike from rotating all the way over and crashing you onto the ground. This actually happened to me once when I was mountain biking with my son and I broke a couple of ribs! I can attest from personal experience that going downhill greatly enhances the likelihood of rotating forward!

The following question illustrates the physics of track and field competitions. In the high jump, the center of gravity of the jumper usually passes under the bar. The same happens for pole-vaulters.

Question: If a high jumper clears the bar, is it possible that the center of mass of the body of the jumper passes below the bar? If so, can you make me visualize the scenario through a video or image illustration or a vivid description? I think that the center of mass can be below the bar during the jump, but it has come there after traveling above the bar.

Answer: You can find dozens of pictures and videos on the internet. A nice one is shown in figure 1.9. The path under the bar of the center of mass of the jumper is shown. When the body is bent the center of mass is outside the body. Going over with the back down is called the Fosbury flop after Dick Fosbury, the American high jumper who won the gold medal at the 1968 Olympics and invented this technique.

1.3 Physics around the house

Physics is helpful for myriad projects and activities around the house. Here are a few examples.

Do you need to move a heavy item into your home and you are nervous about a disaster if it is too much for you? Ask the physicist!

Question: I bought a 400 lb gun cabinet and need to pull it on a two-wheel handcart up a 12 ft ramp at about 35 degrees to the horizontal. How much load do one or two people have to carry and how much is borne by the wheel? I am trying to make sure we can be comfortably safe!

Answer: I could make a rough estimate but would need to know the dimensions of the cabinet and if the center of gravity is near the geometrical center. I assume that the cabinet will be parallel to the ramp when it is being pulled.

Follow-up question: It is $20 \times 29 \times 55$. It will not be parallel to the ramp, but about 20 degrees from the ramp (which is about 35 degrees to the ground (thus avoiding four steps).

Answer: Since only an approximation can reasonably be done here, I will model the case as a uniform thin stick of length L with weight W, normal force N of the incline on the wheel, and a force F which you exert on the upper end. In figure 1.10, I have resolved F into its components parallel (x) and perpendicular (y) to the ramp. Next write the three equations of equilibrium, x and y forces and the torques; this will give you the force you need to apply to move it up the ramp with constant speed.

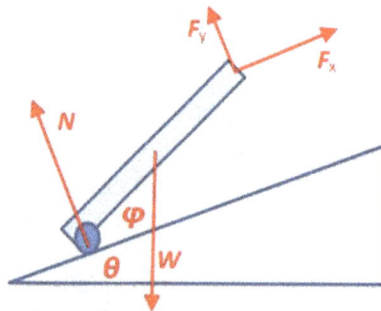

Figure 1.10. Moving a gun cabinet.

$$\Sigma F_x = 0 = F_x - W \sin \theta$$
$$\Sigma F_y = 0 = F_y + N - W \cos \theta$$
$$\Sigma \tau = 0 = \frac{1}{2}WL \cos(\theta + \varphi) - NL \cos(\varphi)$$

I summed the torques about the end where you are pulling. Putting in $W = 400$ lb, $\theta = 35°$, and $\varphi = 20°$, I find $F_x = 229$ lb, $F_y = 206$ lb, and $N = 122$ lb. Note that you do not need to know the length L. The net force you have to exert is $F = \sqrt{[(F_x)^2 + (F_x)^2]} = 308$ lb. If someone were at the wheel pushing up the ramp with a force $\textbf{\textit{B}}$, that would reduce both F_x and F_y. This would change the equations to

$$\Sigma F_x = 0 = F_x - W \sin \theta + B$$
$$\Sigma F_y = 0 = F_y + N - W \cos \theta$$
$$\Sigma \tau = 0 = \frac{1}{2}WL \cos(\theta + \varphi) - NL \cos(\varphi) + BL \sin(\varphi)$$

For example, if $B = 100$ lb, the solutions will be $F_x = 129$ lb, $F_y = 169$ lb, and $N = 159$ lb; so the force at the top will be reduced by about 30% to $F = 213$ lb.

I later received the following message from the questioner: 'Using your info, three of us were confident (and succeeded) in moving the cabinet safely up a ramp over four steps.'

The following question is one I am not likely to get again! Still, the questioner is asking an intelligent question regarding her 'cat highway' project and knows enough physics to appreciate that the force a shelf will feel will be larger than the weight of a cat.

Question: I am building a customizable 'cat highway' of wooden shelves in my living room. The issue I am having is figuring out how much holding force I need, rather than how much weight the actual shelf can support. I know the shelves are plenty strong enough. Now I need to have them fastened strongly enough to the wall. It would be easy, if all I had to consider was the maximum weight a shelf has to take, if all three cats were to lie on it at once. However, I also need to know how much weight I have to budget for impact force—both from a descending and an ascending cat. As there isn't room for multiple cats to jump up or come down at one time, only the weight of the heaviest cat (3.3 kg) needs to be used for the calculations. Also, there will not be a height of more than 0.5 m from one shelf to another. I can't guarantee that they won't skip a shelf, so it might be safer to double the maximum distances just to be safe.

Answer: Normally I answer such questions by saying 'you cannot tell how much force results if an object with some velocity drops onto something'. The reason is that the force depends on how quickly the object stops; that is why it hurts more to drop on the floor than to drop on a mattress. In your case, however, we can estimate the time the cat takes to stop because we can estimate the length of its legs, which is the distance over which it will stop. You are probably not interested in the details, so I will give you the final answer: assuming constant acceleration during the stopping period, the force F necessary to stop a cat of mass m, falling from a height h, and having legs of length ℓ may be approximated as $F = mg(1 + (h/\ell))$. For example, if $h \approx 0.5$ m and $\ell \approx 0.1$ m, $F \approx 6\ mg$, six times the weight of the cat.

Follow-up Question: Thank you for your answer! I thought you might want to know that, unlike a lot of people, I AM actually interested in the details.
Answer: OK, here goes: I will use a coordinate system with increasing y vertically upward and $y = 0$ at the landing shelf. The cat will have acquired some velocity $-v$ when his feet hit the shelf. Assuming he stops after going a distance ℓ and accelerates uniformly, we have the two kinematic equations $0 = \ell - vt + \frac{1}{2}at^2$ and $0 = -v + at$. From the second equation, $t = v/a$; so, from the first equation, $0 = \ell - v(v/a) + \frac{1}{2}a(v/a)^2 = \ell - \frac{1}{2}v^2/a$, so $a = \frac{1}{2}v^2/\ell$. Now, as the cat is landing there are two forces on him, his own weight mg down and the force F of the shelf up, $-mg + F$, and this must be equal, by Newton's second law, to ma, so $F = m(g + \frac{1}{2}v^2/\ell)$. This, by Newton's third law, is the force which the cat exerts down on the shelf. Finally, the speed if dropped from a height h is $v = \sqrt{(2gh)}$, so $F = mg(1 + (h/\ell))$.

The next question is from a mother who would prefer that her children are not crushed by a nearly 500 lb kitchen counter which she plans to have built. *The Physicist* has a serious responsibility as a protector of young children sometimes!

Question: These are non-academic, practical questions regarding the physics of balance and weight shifting pertaining to a moveable kitchen island.
    Q1: Will an 11 inch countertop overhang alone cause the island to tip?
    Q2: If not, what amount of weight can be safely placed on the overhang before tipping?
    Q3: Is there a formula I can use to calculate this?

Description: I have a moveable kitchen island fabricated with locking casters. I wanted to place a quartz countertop over the island base with an additional 11" countertop overhang supported by steel beams extending from the island base. The total length of the countertop is 37" (26" on the island base plus 11" as the overhang).
Properties of the island:
- Back to front length 26"
- Weight of the island base with its 26" portion of the countertop = 414 lb
- Center of gravity, excluding overhang, 14" from the back edge
- Overhang length = 11"
- Weight of the overhang = 58 lb
- The casters are recessed 2" from the front and rear edges

The overhang is a key feature of the kitchen island for our household. The island is intended to serve multiple, mundane purposes in our very compact home: food prep, dining, and working on work projects/homework.

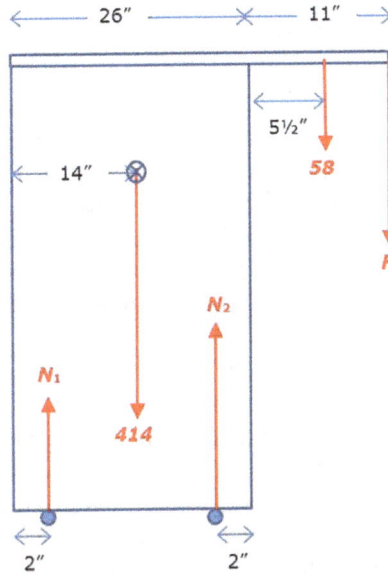

Figure 1.11. A moveable kitchen island.

Answer: The red vectors in figure 1.11 are the pertinent forces for this problem, the 414 lb weight of the island acting at the center of gravity, the 58 lb weight of the overhang acting at the center of gravity (5.5″ out), the force of the floor on the front casters (N_2), and the force of the floor acting on the rear casters (N_1). (*Ignore the force F for now.*) Newton's first law stipulates that the sum of all the forces must be zero, so $N_1 + N_2 = 472$. Also required for equilibrium is that the sum of torques about any axis must be zero; choosing to sum the torques about the front casters, $22N_1 - 414 \times 10 + 58 \times 7.5 = 0 = 22N_1 - 3705$ or $N_1 = 168$ lb, and so $N_2 = 304$ lb. Now, let's think about this answer: it tells you that this (unladen) island will not tip over because there is still a lot of weight on the rear wheels. Now, if you start adding weight to the overhang, eventually when you have added enough weight, the force N_1 will equal zero when it is just about to tip over. So, add the force F at the outermost edge of the overhang and find F when the island is just about to tip: again summing torques about the front casters, $58 \times 7.5 + 13 \times F - 414 \times 10 = 0$ or $F = 285$ lb. This is the extreme situation—you would have to put nearly 500 lb, for example, halfway out on the overhang to tip it over. It looks to me like this will be safe for everyday use.

Gardeners need physics too. In the following question a gardener plans to get his garden started early in the season by planting seeds in a cold frame, a glass enclosure to protect against the elements. His worry, though, is that he lives in a locale where high winds are frequent; he worries that the hinged doors might be lifted by the Bernoulli effect—high-velocity air causes lowered pressure. We talked briefly about this effect in section 1.2 in the context of baseballs, but examine it in more detail here.

Question: I am building a cold frame to keep veggies alive in the winter. It will be 3′ × 6′ with two 3′ × 3′ 'doors' (called lights) that will be hinged to the frame. The frame probably weighs about 10–15 lbs. The doors will be quite lightweight, possibly only 3 lbs each. I wanted to use magnets to keep the doors closed at night or when I am not venting the cold frame. We often get very windy days with 40 mph wind speeds and gusts to 60 mph. From what I've read, magnets have different 'pull force' properties. I'd like a way to figure out what pull force the magnet for each door needs to have to withstand the winds we get. Please don't tell me to just hook the doors closed—shockingly, it appears that magnets are a more cost-effective solution.

Answer: I must say that I cannot believe that you could not buy a couple of simple hooks/latches for under $5, but I will do a rough calculation for you to estimate the force you would need to apply at the edge of the doors to hold them down in a 60 mph = 26.8 m s^{-1} wind. When a fluid moves with some speed v across a surface, the pressure is lower than if it were not moving; this is how an airplane wing works and why cigarette smoke is drawn out of the cracked window of a moving car. To estimate the effect, Bernoulli's equation is used: $\frac{1}{2}\rho v^2 + \rho g h + P = $ constant, where ρ is the density of the fluid ($\rho_{air} \approx 1$ kg m^{-3}), P is the pressure, g is the acceleration due to gravity, and h is the height relative to some chosen $h = 0$. For your situation both surfaces are at essentially the same height, so $P_A + \frac{1}{2} \times 1 \times 0^2 = P_A = P + \frac{1}{2}\rho v^2$, where the pressure inside your frame is atmospheric pressure (P_A) and the velocity inside is zero. So, $P_A - P = \Delta P = \frac{1}{2}\rho v^2 = \frac{1}{2} \cdot 1 \cdot 26.8^2 = 359$ N m^{-2} = 7.5 lb ft^{-2}. This would be the pressure trying to lift the door. So the total force on each door would be $F = A\Delta P = 9 \times 7.5 = 67.5$ lb, where $A = 3 \times 3$ ft^2 = 9 ft^2 is the area of the door. But, this is not the answer to your question because we want to keep it from swinging about the hinges, not lifting into the air. So, assuming that the force is distributed uniformly over the whole area, you may take the whole force to act in the middle, 1.5 ft from the hinges, so the torque which is exerted is $1.5 \times 67.5 \approx 100$ ft · lb. However, the weight of the door also exerts a torque, but opposite the torque due to the wind (the weight tries to hold it down) $-3 \times 1.5 = -4.5$ ft · lb. So, the net torque on the door about the hinges is about 95 ft · lb. To hold the door closed, one needs to exert a torque equal and opposite to this. To do this, it would be wisest to apply the force at the edge opposite the hinges to get the maximum torque for the force. The required force from your magnets would then be $F = 95/3 = 32$ lb. Note that this is just an estimate. Fluid dynamics in the real world can be very complex. Also note that, if my calculations are anywhere close to correct, you should probably be sure the whole thing is attached to the ground or the side of your house since the total force on the whole thing would be $67.6 + 67.5 - 3 - 3 - 15 = 114$ lb, enough to blow the whole thing away in a 60 mph wind! Also, once the door just barely opens, the wind will get under it and simply blow it up, Bernoulli no longer makes any difference.

This next question is really pretty simple, but uses common sense and the notions of center of mass and Newton's first law to get a good estimate of the weight of a very heavy object.

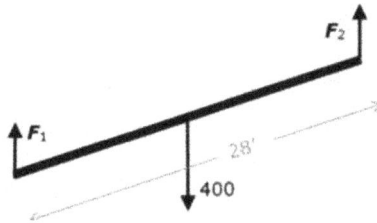

Figure 1.12. A ramp.

Question: We have a kitchen table that is extremely heavy and we know of no way to weigh it. If I put one leg on a scale, it reads 143.2 lbs. I'm thinking we can't simply multiply that number by four because it seems it may register some of the weight from at least part of the other three quarters.
Answer: Get a 2 × 4 which will span the two legs at one end and weigh it; call that weight w. Put the 2 × 4 under the legs at one end and the scale under that; the scale will read W_1. Move the 2 × 4 and scale to the legs at the other end; the scale will read W_2. The weight of your table is $W = W_1 + W_2 - 2w$. If your table is symmetrical, you will find W_1 to be about the same as W_2.

Next is another pretty straightforward question. The questioner was surprised that it makes no difference what the rise angle is.

Question: If I have a ramp that is 28 feet long, is fixed at the upper end (shore), weighs 400 lbs and has a six-foot rise, how do I calculate the weight at the lower end (dock)? I am trying to determine how much flotation I need under the water end to support the weight of the ramp at that end. That rise varies during the course of the year from zero feet to a maximum of seven feet (see figure 1.12).
Answer: I am assuming that the ramp is a uniform plank, that is, that its center of gravity is at its geometrical center (14'). Refer to figure 1.12. Two equations must be satisfied for equilibrium, the sum of all forces must equal to zero and the sum of all torques about any axis must be zero. The first condition gives us that $F_1 + F_2 - 400 = 0$ and the second condition (summing torques about the center of the ramp) gives us that $F_1 - F_2 = 0$. Solving these two equations, $F_1 = F_2 = 200$ lb. Note that the answer, 200 lb, is independent of the rise.

1.4 How does that thing work?

Things go on around us that are sometimes puzzling. How does a device work? How can I improve the performance of my product? Why is that thing designed the way it is? Is somebody else's product design superior mine? *Ask the Physicist*!

Question: When I had solar panels installed on my house the old-fashioned meter ran backwards when the Sun was bright. They have now fitted a digital meter which can sense when energy is being sent into the grid. I can see how this could be done with DC, but how does it work with AC? In AC the current is switching direction at 50 Hz. How can the meter sense the 'direction' of the energy flow? **Answer:** You are right, the average current is zero. However, the current is not the power—the power is the product of the current and the voltage. Both current and voltage are sinusoidal functions of time, $i(t) = I\sin(\omega t)$ and $v(t) = V\sin(\omega t + \varphi)$, so $p(t) = IV\sin(\omega t)\sin(\omega t + \varphi)$. Figure 1.13 shows three choices for the phase φ between i and v. For $\varphi = \pi/2$, the time average of the power is zero, no energy flow; for $\varphi = \pi$ and $\varphi = 0$, the time average of the power is negative and positive, respectively. The motor in a mechanical meter turns in opposite directions for different signs of the average power; in a digital meter the average power is determined by an electronic circuit.

The next three questions are all about Archimedes' principle, which might be the oldest physics principle, more than two millennia old, that is actually correct. Simply put, the principle states that when an object is in a fluid it feels an upward force, called the buoyant force, which is equal to the weight of the fluid which it displaced. Because of Archimedes' principle, boats float and hot air balloons rise. First, a short answer which explains the origin of the buoyancy, and then a second question with an application of floating the very heavy lighthouse lens in a pool of mercury.

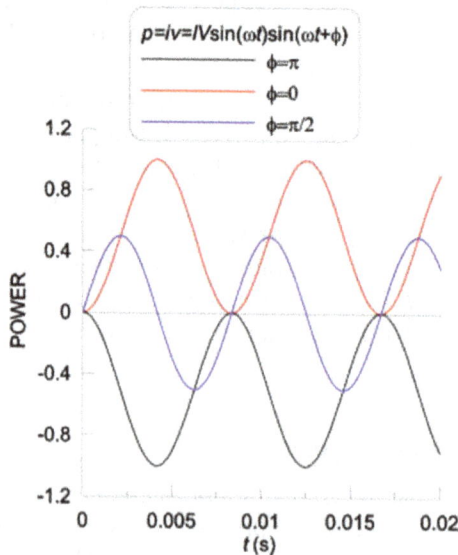

Figure 1.13. Power in an AC circuit.

Question: I hope this question is stupid and has a very quick answer. If I have an object submerged in a tank of water, say a solid cylinder, obviously the pressure at the top of the cylinder will be less than the pressure at the bottom of the cylinder, due to the difference in depth between the two points ($P = \rho g h$). What I would like to know is how this difference in pressure exerts a force on the cylinder, and where? Is the force acting on the bottom of the cylinder, trying to push it to the surface? I just can't wrap my head around this at all.

Answer: Sorry to disappoint you, but your question is not stupid. You have discovered Archimedes' principle! Since, as you note, the pressure at the bottom of the object (pushing up) is greater than the pressure at the top of the object (pushing down), there is a net push up. This force is called the buoyant force. It may be shown that the buoyant force is equal to the weight of the water displaced by the object and always points up. The buoyant force determines whether an object floats (buoyant force greater than weight) or sinks (buoyant force smaller than weight) and why swimming is sort of like flying.

Question: I am a volunteer guide at South Foreland historic lighthouse in the UK. We have an optic weighing approximately two tons, floating in a close-fitting trough containing approximately 28 liters of mercury. What is the theory which enables this optic to float, as it does not appear to fit within the basics of Archimedes' principle?

Answer: To float two metric tons (2000 kg) you must displace $M = 2000$ kg of mercury. The density of mercury is $\rho = 13\ 600$ kg m^{-3}, so the volume you must displace is $V = M/\rho = 0.15$ m$^3 = 150$ l; this, I assume, is what is bothering you, since only 28 l are used. Suppose that the reservoir for the mercury is a cylinder of radius 1 m and depth d; to contain 0.15 m^3, the depth of the container would have to be $d = 0.05$ m $= 5$ cm (estimating $\pi \approx 3$). So, I will make a container 6 cm high for an extra 20%, 180 l, and I will buy that 180 l to fill it up. Now, let's make the pedestal on which the lens sits be a solid cylinder of radius 99 cm, so that when you put it into the reservoir there will be 1 cm gap all around. So as you lower it into the reservoir, mercury will spill out the top and you will be sure to capture it. When you have captured 150 l, the whole thing will be floating on the mercury. You return the extra 150 l and have floated the lens with only $180 - 150 = 30$ l. Of course, in the real world you would only buy 30 l, put the pedestal into the empty reservoir, and add mercury until it floats. (I realize that the shapes of the pedestal and reservoir are probably not full cylinders, since you said 'trough', but my simple example wouldn't be so simple with more complicated volumes. The idea is the same, though.)

Speaking of Archimedes' principle, this next question supposedly extends it to an avalanche-escape product which inflates and pops you to the surface, assuming the snow will behave like a fluid.

Question: I have just started a new job and have been asked to do a presentation on the ABS (Avalanche Bag system). Upon reading about the product, they kept quoting that the system works on the 'law of inverse particles'. I figure ok, it's a law, it should be fairly simple to research, but when I started typing it into search engines I got no feedback on this law at all. So my question is does this law even exist? If so, is there maybe another name that it goes by? If it does exist, who created it? Any help that you can offer me on this topic will be very much appreciated, in the meantime I'll keep on checking books, sites and asking old teachers!

Answer: I have never heard of the law of inverse particles. However, just looking over how this thing is supposed to work, it just looks like Archimedes' principle to me. Archimedes' principle says that when an object is placed in a fluid it experiences an upward force, called the buoyant force, which is equal to the magnitude of the weight of the displaced fluid. Hence, if the object has a density less than the density of the fluid, it will float, if it is greater it will sink. I am guessing that the snow in an avalanche acts much like a fluid and, if you can significantly decrease your overall density, you will 'float' on the surface, much like a cork floats on water. The way you decrease your density is to increase your volume with the inflated airbags. Why they invented this screwy name for this escapes me—it must be an effective marketing technique.

The next two questions are about commercial products which have encountered problems related to their performance in the marketplace.

Question: As part of our business we bag wrap passengers' bags and suitcases prior to flying at the major UK airports. We use and have used for many years a power pre-stretch cast film 17 micron nano with a 300% capability. Recent feedback from Heathrow airport suggests some of the passengers' bags are sticking to the conveyer belts and are being misdirected. I am being asked for the 'coefficient of friction' (COF) for the film we are using. I have advised our supplier of this, they have sent through the data spec sheet, but there is no mention of COF; on speaking with them, they have never had this question raised before. Personally, I do not think this is an issue with our film, but more where customers themselves are wrapping their own bags with home-use film. However, I need to provide proof that the film we are using does not have any adhesive properties. My question is—would the COF affect this and how do I get the actual information on the film?

Answer: The force of friction f depends on only two things: what the surfaces which are sliding on each other are (conveyer material and your plastic film) and the force N which presses the two surfaces together. Normally, on a level surface, the force N is simply the weight of the object (a suitcase here). There are two kinds of COFs, kinetic and static. The kinetic coefficient, μ_k, allows you to determine the frictional force on objects which are sliding. In that case, $f = \mu_k N$. The static

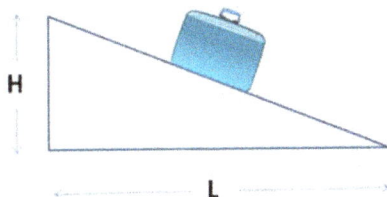

Figure 1.14. Measuring the static coefficient of friction.

coefficient, μ_s, allows you to determine how hard you have to push on the suitcase in order for it to start sliding; in this case $f_{max} = \mu_s N$, where f_{max} is the greatest frictional force you can get. Since you are being asked to prove that it is not too 'sticky', it is the static, not the kinetic, coefficient which you need; measuring μ_s is quite easy. The only problem is that μ_s depends on the surfaces, so you must have a piece of the material from which the conveyer belts are made to make a measurement. Once you have that, use it as an incline on which to place a wrapped suitcase. Slowly increase the slope of the incline until the suitcase just begins to slide. Your COF (μ_s) is equal to the tangent of the angle of the incline which (see figure 1.14) is simply $\mu_s = H/L$.

Question: In the old days, many types of tags were made out of paper. Nowadays, lots of tags are made out of film, because the film is waterproof, grease resistant, and stronger than paper. One application where film tags are being used is in the grocery store for items like hams, turkey, and fish. Our company, YUPO, makes film that is used in grocery stores to tag hams, but unfortunately, the grocery store is complaining because the tags are failing in the store. I am writing to ask for some help in determining how much strength is required from my film to actually work for this application. I hope you can help. Now for the set-up to the question. Customers pick up the ham using the tag. They raise it to about shoulder height. They lower it into the cart holding onto the tag. Then, suddenly, at the last second, they jerk the tag upwards so the ham doesn't crash into the other items in the cart. When the customer jerks it up at the last second, the tag snaps. The hams weigh 10 pounds. I am wondering how much force is required to prevent the tag from snapping. I think this can be explained through physics, but I don't know how to do it. Our lab has equipment that can be used to test tensile strengths, elongation, etc, but I am asking for your help in figuring out how much strength will be required. Can you help me?
Answer: Every customer is going to lower it differently, so it is, of course, impossible to give a definitive answer. I have worked out the force F which the neck would have to exert on an object of weight W (lb) if the customer simply dropped the object from a height h (ft) and then stopped it completely in a distance s (in): $F = W[1 + 1.06\sqrt{(h/s^2)}]$. For example, if a 10 lb ham dropped from 2 ft and stopped in 2 in, $F = 17.5$ lb. That would be a pretty extreme case, though, since I would guess most people would probably lower it at a lower speed than it

would achieve in free-falling 2 ft. I think engineers like to insert a factor of two safety factor. Overall, I would guess that the tag should to be able to handle at least twice the weight of the product.

The next question requires using Bernoulli's equation for fluid dynamics to determine the rate of fluid flow in a hospital feeding device for neonatal babies.

Question: If I have two hoses of varying diameters, such as a garden hose and a fire hose, that are vertical and approximately 20 feet long, and are filled with the same amount of water and have the same size opening at the bottom, using just gravity, will water flow through each hose in the same amount of time? This question came up during a recent visit with a customer. I sell feeding supplies for neonatal patients. The current product being used is a large bore tube compared to the smaller bore that my company sells. However, the opening at the distal tip of the feeding tube is the same size for both. Again, using just gravity, shouldn't the speed at which formula flows through the distal end be the same since it bottlenecks there?

Answer: Your first question ('will water flow through each hose in the same amount of time?') is ambiguous, so let me answer the question by finding how the speed of the delivered formula (labeled V_{bottom} in figure 1.15) depends on the geometry. The operative physics principle is Bernoulli's equation, $P + \frac{1}{2}\rho V^2 + \rho g y = $ constant, where P is pressure, ρ is the fluid density, y is the height above some chosen reference level, V is the fluid velocity, and $g = 9.8$ m s^{-2} is the acceleration due to gravity. In your case, P is atmospheric pressure at both the top and the bottom, and I will choose $y = 0$ at the bottom, so $y = h$ at the top.

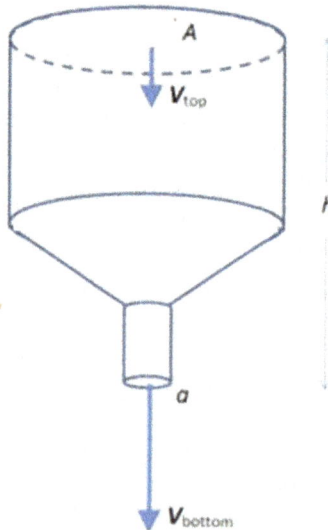

Figure 1.15. Feeding tube for neonatal babies

Figure 1.16. Pipeline expansion loops.

Therefore Bernoulli's equation becomes $V_{top}^2 + 2gh = V_{bottom}^2$ or $V_{bottom} = \sqrt{(2gh + V_{top}^2)}$. There are two ways that V_{bottom} will be independent of the geometry: (1) if h is held constant by replenishing the formula at the top or (2) if the area of the bore is much larger than the area of the distal tip $A \gg a$. Both of these result in $V_{top} \approx 0$, so $V_{bottom} \approx \sqrt{(2gh)}$. If neither of these is true, it is a much more complicated problem.

Pipelines are often in the news. Often protests center around the possibility of pipeline failure. One problem that has to be dealt with is the possibility of rupture due to thermal expansion and contraction as ambient temperatures vary. The next question demonstrates how this is handled.

Question: Why are loops provided for transporting oil/water for longer distances?
Answer: When the temperature of the pipe changes, it changes length. In figure 1.16, the pipe will expand if you heat it up and contract if you cool it down. If it were just a straight pipe, the resulting forces on the pipe along its length could be large enough to cause it to buckle and fail. Inserting loops allows the length changes to be taken up by the size of the loop.

Chapter 2

Physics is beautiful

2.1 Introduction

'Beauty lies in the eyes of the beholder' (Plato). What is beautiful? Art, music, literature, a well-executed football play, a really great bottle of wine? These are among the things that many people consider beautiful, but they would certainly not be universally proclaimed to be beautiful. My wife has no appreciation of football and wine is not my cup of tea. The one thing which I believe would have the most votes for beauty is nature—a sunset, the view from a mountain top, a virgin forest, a rainbow, a snow-covered mountain range, a clear night with a star-filled sky, a vast desert at dawn. Science is the examination of nature, good science seeks to examine the beauty of nature. To a physicist, there is beauty not just in beholding a sunset or a rainbow or the Milky Way, but also in understanding more deeply how such wonders can be. Physicists seek to uncover laws which are simple, compact, and universal; mystery may be one face of beauty, but understanding is another form of beauty to a scientist—elegance is the ultimate beauty of understanding. In this chapter I will select questions which ask for an understanding of nature. Whether the answers reveal elegance and beauty will be for the reader to judge.

2.2 The beauty of electromagnetism

Of all the fields of physics, the theory of electricity and magnetism (electromagnetism) is arguably the most beautiful. It is physics' most complete and best understood theory. It has been the foundational motivator for other beautiful theories, for example:
- quantum mechanics (black body radiation, Compton scattering, the photoelectric effect);
- special relativity (transformation properties of electric and magnetic fields, the speed of light);
- optics (electromagnetic radiation).

The mathematical structure of the theory is amazingly compact, all aspects apart from the quantization of the field are contained in four concise equations (Maxwell's equations); in fact, using the notation of special relativity, the four reduce to only two.

This beautiful theory, however, was hard-won. Although electrical and magnetic forces were known in antiquity, little scientific understanding was achieved before the 17th century; it was not fully developed until the middle of the 20th century, when the theory was quantized, after about 350 years.

The earliest serious scientific study of electromagnetism was by William Gilbert in 1600. He was the first to appreciate that electricity and magnetism are not the same thing (although we later came to realize that they were both aspects of a more general electromagnetic field). Nearly 200 years passed before a solid foundation for electrostatics (electric charges at rest) was published by Charles Augustin de Coulomb in 1794; Coulomb's law tells us the force between two point charges.

Question: Why do we divide by r^2 in Coulomb's law and what is the role of 4π in the constant of proportionality in the law? I am not able to understand why we divide the product of q_1q_2 by $4\pi r^2$.

Answer: The reasons for r^2 and 4π are different. Coulomb's law is a statement of an experimental fact. If you have two charges, q_1 and q_2, and measure the force F they exert on each other and then double either charge, the new force will be twice as great; you have therefore found out *experimentally* that $F \propto q_1$ and $F \propto q_2$. Now, if you keep the charges constant and double the distance between them, you will find that the force gets four times smaller; you have therefore found out *experimentally* that $F \propto 1/r^2$. Putting it all together, Coulomb's law tells you that $F \propto q_1q_2/r^2$. But we usually prefer to work with equations rather than proportionalities, so we introduce a proportionality constant k: $F = kq_1q_2/r^2$. The usual way to determine k is to measure F for a particular q_1, q_2, and r. (Note that the SI unit of charge, the coulomb (C), is defined independently of Coulomb's law; it is defined in terms of the unit of current, the ampere (A), 1 C/s = 1 A.) You find that $k = 9 \times 10^9$ N $\cdot$ m^2 C^{-2}. Another way to put it is that you would find that two 1 C charges separated by 1 m will exert a force of 9×10^9 N on each other. That answers your first question about why the $1/r^2$ appears in Coulomb's law—it is simply an experimental fact, it is the way nature is. Your second question is why do we often see the proportionality constant written as $k = 1/(4\pi\varepsilon_0)$. There is nothing profound here; later on, when electromagnetic theory is developed further, choosing this different form leads to more compact equations. Essentially, many equations involve the area of a sphere, which is $4\pi r^2$, which means that there would be many factors of 4π floating around in your equations of electromagnetism if you used k as the proportionality constant.

The similarity of Coulomb's law, $F = kq_1q_2/r^2$, to Newton's universal law of gravitation, $F = Gm_1m_2/r^2$, is striking. As alluded to above, the $1/r^2$ has its origins in the fact that a sphere surrounding a mass or a charge has an area of $4\pi r^2$; think of the force as being 'spread out' over the whole sphere, so its strength gets 'diluted' the farther you get from the source. They are, though, strikingly dissimilar in magnitude: in SI units, the force between two 1 kg (1 C) masses (charges) separated by 1 m is 6.67×10^{-11} N (9×10^9 N)—20 orders of magnitude! Gravity is nature's weakest force; it only seems like the strongest because we happen to be in the vicinity of a huge amount of mass. Another striking difference is that there is only one kind of mass, but there are two different kinds of electric charge. This reveals itself in the gravitational force being always attractive, whereas the electrostatic force can be either attractive (between different kinds of charges) or repulsive (between the same kinds of charges). The two kinds of charges were first denoted as positive and negative by Benjamin Franklin in the 18th century.

Magnetism was thought to be an entirely different phenomenon from electricity until early in the 19th century. It was believed to be present only in certain minerals and its principle application, the compass, played a vital role in civilization; Gilbert had proposed that compasses worked because the Earth itself is a magnet. Then, in 1820, a Danish physicist, Hans Christian Ørsted, made an important discovery. While setting up demonstrations for a university class which involved both electric currents and several compasses, he noticed, out of the corner of his eye, that when the current started to flow the compasses moved.

Question: Is there a metaphor or a simplistic explanation that I can use to explain to middle schoolers that electricity and magnetism are manifestations of the same force? Age-appropriate textbooks just state that this is so without explanation, or else they simply give the example of an electromagnet, which doesn't answer the deeper question. Some books don't even explain that they are related! Introducing the ideas of vectors and tensors makes the explanation too long and I lose most kids' attention. Do you have any ideas?

Answer: Sure, there are lots of ways to demonstrate the connection. Easiest is to simply refer to an electromagnet. Here we have an electric current going around a coil of wire and it will attract iron. You can easily make one by wrapping wire around a nail and connecting it to a battery and show that the nail becomes a magnet. Moving electric charges (which is what the current in the wire is) cause a magnetic field. This connection was first discovered by Ørsted, a Danish physicist of the 19th century, who noticed that, when he hooked up an electrical circuit, a nearby compass deflected from north. You can also demonstrate the connection by changing a magnetic field, for example by thrusting a magnet into a coil of wire; the changing magnetic field causes an electric current to flow in the coil.

It is not possible to overstate the importance of this discovery. Knowing that electric currents cause magnetic forces opened up the possibility of electric motors and electric generators, devices imperative to the technological progress of the time. And, any development in science which unites two branches previously thought to be unrelated is an advance for beauty and elegance.

With electricity and magnetism now united, progress in the 19th century accelerated. In the first half of the century a brilliant British physicist, Michael Faraday, made groundbreaking contributions. In my view, his most important contribution was an idea which was not accepted by the scientific community until after his death. He proposed the idea of electric and magnetic fields, a representation of what is contained in the space around a force without making reference to any specific forces. For example, the electric field E of a point charge q at a distance r is $E = kq/r^2$. Knowing the field allows us to infer the force F felt by any other charge q' brought into the region—$F = q'E$. The magnetic field is notated as B. This may seem like an artificial construct, but it turns out that the space around charges or currents is physically altered—it can be shown that the space has energy content. Faraday's other major contribution was what is now called Faraday's law; basically, the law says that a changing magnetic field causes an electric field to appear. Faraday was an amazing fellow. He had almost no formal schooling and was almost entirely self-educated. Starting at age 14 he was apprenticed to a bookbinder, where he stayed for seven years. During those seven years he read voraciously and acquired an interest in physics and chemistry. He was fortunate to then get a position with the eminent chemist Sir Humphry Davy as a laboratory assistant. Through perseverance and hard work, he became one of the most brilliant and accomplished physicists of the 19th century.

In the second half of the 19th century, much was known about electricity and magnetism. But what was known was not really a theory, rather it was a compendium of experimental facts; it was not beautiful! This final beautification fell to James Clerk Maxwell. He was able to reduce all that was known about electromagnetism to four equations, now known as Maxwell's equations.

Question: I'm trying to understand Maxwell's equations, but I can't! There is a lot of maths and I'm only 15, which means I'm not ready for that kind of level! But how can I be a physicist without understanding Maxwell's equations?
Answer: Since you cannot do the mathematics, I can only give you a brief qualitative explanation of what Maxwell's four equations say. Maxwell's equations tell you everything there is to know about electricity and magnetism. They also contain the theory of special relativity; they predict the speed of light and demonstrate that light is nothing more than varying electric and magnetic fields. You need to know what an electric field and a magnetic field are: electric fields, when present, cause an electric charge to feel a force and magnetic fields cause a moving electric charge to feel a force. Here is a qualitative list of all the contents of Maxwell's equations:
- electric charges cause electric fields;
- electric currents (moving charges) cause magnetic fields;
- changing electric fields cause magnetic fields;
- changing magnetic fields cause electric fields;
- magnetic monopoles (magnetic charges) do not exist.

That is about all you can do without math.

It is interesting that Maxwell did more than just put together what was known; when he had his four equations written, he noted that although there was a term indicating that a changing magnetic field would cause an electric field (Faraday's law), there was no comparable term for a magnetic field being created by a changing electric field. He invented such a term and inserted it purely on the basis of aesthetics; it is natural for a good scientist to seek the most elegant, beautiful, and concise theory! Electromagnetism is now beautiful.

But, there is more, maybe the best, yet to come.

Question: Why is the speed of light given by $1/\sqrt{(\varepsilon_0\mu_0)}$? What is the great mystery behind such a simple relation? How do these two parameters combine to give the speed of light? Why does the vacuum (nothing) have physical properties such as permittivity and permeability?

Answer: This is the great triumph of Maxwell's work in the 19th century. There are laws of electromagnetism, which can be summarized in four equations, now known collectively as Maxwell's equations. The quantity ε_0 (permittivity of free space) is just a proportionality constant which tells you how strong the electric force is and, of course, it appears in the equations. Similarly, the quantity μ_0 (permeability of free space) is just a proportionality constant which tells you how strong the magnetic force is and, of course, it appears in the equations. (In this context, there is nothing wrong with empty space having permittivity and permeability because one certainly does not need matter between charges or currents for them to exert forces on each other.) When Maxwell messed around with the equations he discovered that they could be rewritten as wave equations and that the speed of these waves had to be $1/\sqrt{(\varepsilon_0\mu_0)}$. That this happened to be the speed of light meant that this was the point in the history of physics where we understood what was doing the waving in light waves—electric and magnetic fields. (See appendix A for a more detailed discussion of the permittivity and permeability constants; see appendix B for a derivation of the wave equation.)

What is not apparent in this answer is how significant a discovery this was. Early in the 19th century much was known about light. Its speed had first been measured by the Danish physicist Ole Rømer in 1676 using a brilliant method that involved observing eclipses of the Moons of Jupiter and was extremely accurately known to be $c = 3 \times 10^8$ m s^{-1} by the mid-19th century. In 1803 the British physician Thomas Young also clearly demonstrated that light was a wave; but no one knew what was 'waving'. That waves should be predicted by Maxwell's equations to have a speed of $1/\sqrt{(\varepsilon_0\mu_0)} = 3 \times 10^8$ m s^{-1} could not possibly have been a coincidence. Figure 2.1 shows a representation of how we think of electromagnetic waves today. A derivation of the wave equation from Maxwell's equations may be found in appendix B.

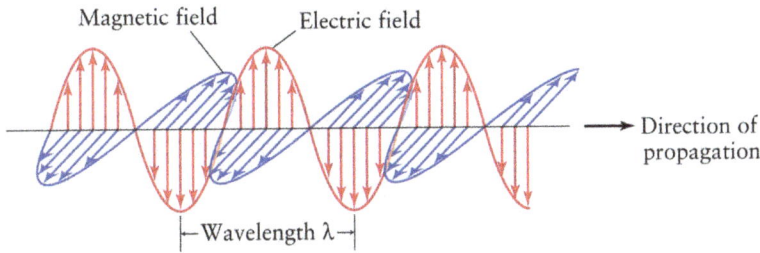

Figure 2.1. An electromagnetic wave.

2.3 The beauty of relativity theory

There was something odd about Maxwell's equations. Since they were based on experiments, they were the laws of electromagnetism that applied here on Earth, where the experiments were performed. But suppose we wanted to know what the laws of electromagnetism would be in another frame of reference, say an airplane flying overhead at a speed of 1000 mph. We could redo all the experiments leading to Maxwell's equations in the airplane to get the laws of electromagnetism aboard the airplane, but that would be a great nuisance.

Why not simply do a simple transformation (called a Galilean transformation) from our frame of reference to the airplane's frame to get the new laws there? It is an easy thing to do and this kind of thing had already been done for Newton's laws of classical mechanics; see appendix C to see how to transform velocity and acceleration into a moving frame. For example, Newton's second law is $F = ma$, the relation among the force F, mass m, and acceleration a of m which F causes; doing the transformation, we find that the acceleration seen by an observer in the airplane is exactly the same as the acceleration you see on the ground. If this were not the case, this would not really be a useful law of physics, would it? If the acceleration measured in the airplane were twice what you measured from the ground and you deduced that it was being caused by a 1 lb force, the traveler would say it was caused by a 2 lb force. Such a contradiction should tell us that Newton's second law is worthless, not an acceptable law of physics. The fact that this does not happen is part of what makes Newton's second law a beautiful law. Performing the transformation on Maxwell's equations gives a different set of equations altogether and impossible contradictions arise. All of a sudden, our theory is not as beautiful as we had surmised in section 2.2!

The Dutch physicist Hendrik Lorentz discovered that if you used a different transformation from the Galilean transformation, Maxwell's equations would transform to the same form as they had in the stationary frame; but this was purely empirical and was only useful because it worked, while it didn't make any sense—certainly not a beautiful transformation! Albert Einstein approached the problem from a different perspective, a philosophical point of view. He posited that for anything to be a true law of physics, it must be the same in all frames of reference—this is called the *principle of relativity*. So, what does imposing the principle of relativity on Maxwell's equations lead to?

Question: Why exactly does the speed of light come into play in the theory of special relativity? It seems to be rather arbitrary in the sense that the Lorentz transformation would (from what I can gather in my admittedly novice research) yield the same results using any speed that is accepted as being constant, whether it be the speed of light, sound, or the 4:17 to Charleston. Is light specifically used for any reason other than its bearing on optics?

Answer: I know the postulate that c is a universal constant seems very ad hoc. In fact, that postulate is not needed. Here is the reasoning: the cornerstone of special relativity is that the laws of physics must be the same in all inertial frames of reference. Maxwell's equations, the laws of electromagnetism, predict the speed of light to be exactly what it is and to depend only on two fundamental constants of nature (constants which quantify the strengths of electric and magnetic forces). Since Maxwell's equations are laws of physics, they must be the same in all frames of reference and hence the speed of light must be the same in all frames of reference.

This is an earth-shattering conclusion—for the laws of electromagnetism to be the same in all frames of reference, the speed of light must be the same in all frames of reference. Think about it: if you see light coming from a source and measure its speed, you will determine its speed to be c. If you now find some way to move toward the source with a speed $\frac{1}{2}c$, you will measure the speed to be c, not $1\frac{1}{2}c$! Pretty unsettling, but true. Einstein now took this requirement that all observers see the same speed of light and asked what its implications are—he found a new transformation between different frames and, guess what? This requirement results in the Lorentz transformation between frames of reference. Today we still call it the Lorentz transformation. (Personal comment: you have heard of six degrees of separation, the idea that there are at most six links between any two people? In my case, a professor of mine, Otto LaPorte, best known for the LaPorte selection rule, was a student of Lorentz; I am linked by one connection to Lorentz!)

We shall have more to say about relativity in a later chapter—it was a theory which radically changed how we view the ideas of space and time. For the present, though, its connections to electromagnetism and its introduction of the principle of relativity qualify it as a beautiful theory.

As a final addendum to this section, let us look how the Lorentz transformation transforms electric and magnetic fields. The fact that the electric and magnetic fields in the moving frame both depend on both fields in the stationary frame emphasizes the idea that there really is just one field—the electromagnetic field.

Question: The magnetic force F on a charge q moving with velocity v is $F = qv \times B$. If I observe this charge from a car moving with the same speed and direction as q, then its velocity as observed by me will be zero, so the force will be zero. I am not able to understand this dilemma—in one frame the force is nonzero and in the other it is zero.

Answer: The force is not zero in the moving frame. The problem is that the electric and magnetic fields in one frame of reference are not the same as in another moving frame. (This is special relativity.) In your case you first start with a nonzero magnetic field and a zero electric field. Suppose that the magnetic field is in the y-direction, $\boldsymbol{B} = \boldsymbol{1}_y B$, $\boldsymbol{E} = 0$, and the velocity of q is in the x-direction, $\boldsymbol{v} = \boldsymbol{1}_x v$. Then the force will be $\boldsymbol{F} = \boldsymbol{1}_z q v B$ in the z-direction. In the moving frame the new fields will be $\boldsymbol{B}' = \boldsymbol{1}_y \gamma B$ and $\boldsymbol{E}' = \boldsymbol{1}_z \gamma v B$, where $\gamma = 1/\sqrt{(1-(v/c)^2)}$. Note that $\boldsymbol{E}' = \boldsymbol{v} \times \boldsymbol{B}'$; therefore the force, as seen in the moving frame, is $\boldsymbol{F}' = q\boldsymbol{E}' + q\boldsymbol{v} \times \boldsymbol{B}' = \boldsymbol{1}_z \gamma q v B$. Note, however, that $\boldsymbol{F}' \neq \boldsymbol{F}$, they differ by a factor of γ; this is because force is said to not be *Lorentz invariant* and it is not really a useful quantity in relativity. (The vectors $\boldsymbol{1}_i$ are unit vectors in the $i = x$-, y-, and z-directions.)

It is important to note that the force is not the same in both frames. This is because, as noted above, Newtonian mechanics transforms correctly using a Galilean transformation; it therefore cannot transform correctly using another transformation. This tells us that the laws of Newtonian mechanics are not correct laws of physics. In spite of the fact that I would count Newtonian mechanics among the beautiful laws of physics, they still must be considered to be approximate laws; since the forces differ by a factor of $1/\sqrt{(1 - (v/c)^2)}$, we can say that they are beautiful only if v is very much smaller than c, so that $\gamma \approx 1$. Maxwell's equations have led us to expect major changes in the way we understand nature.

2.4 The beauty of gravitational theory

I would be remiss if I did not include the very first beautiful theory of physics, the theory of gravity.

In the 16th century the Danish astronomer Tycho Brahe made amazingly precise measurements of the motion of the planets through the sky; the accuracy was nearly an order of magnitude better than previous measurements. During the last two years of his life, 1600–01, he had a young assistant named Johannes Kepler. After Brahe's death, Kepler began careful analysis of the data from the years of careful observation. His conclusions, published during the second decade of the 17th century, are summarized by his three laws.

- The first law states that the orbit of a planet is an ellipse with a semimajor axis a and with the Sun at one focus of the ellipse.
- The second law states that a planet in its orbit sweeps out equal areas in equal times, so it moves faster as it gets closer to the Sun.
- The third law states that the square of the period T of an orbit is proportional to the cube of its semimajor axis, $T^2 \propto a^3$ or $T^2 = Ka^3$.

Note that these are not really 'laws' of physics, because they are completely empirical—they represent fits to data, not any physics. Kepler pondered what kept the planets moving—keep in mind that he predated Newton, so it was still thought that a push was needed to keep something moving—and he hypothesized that angels were pushing them around. What we now understand in terms of Newton's first law is that we need to find the force which keeps them from flying out in straight line paths. That would have to wait until Newton published his work later in the century. Nevertheless, since physics is, at heart, an experimental science, I would submit that Brahe's and Kepler's work was beautiful.

In 1686 Isaac Newton showed that an inverse square law force between two points or spherically symmetric masses ($F \propto 1/r^2$, with r being the distance between two masses) would predict Kepler's three laws. He also reasoned that the force should be proportional to the product of the masses, m_1 and m_2. Therefore, $F \propto m_1 m_2/r^2$ or $F = Gm_1m_2/r^2$, where $G = 6.67 \times 10^{-11}$ N $\cdot$ m^2 kg^{-2} is a proportionality constant and is called the universal gravitational constant. It does not take very long to realize that this law applies not only to the planet–Sun systems, but also to any two-body pairs—comets which have highly eccentric orbits and the Sun, satellites orbiting planets, binary stars, etc. Interestingly, Newton could not extract G from the data; an accurate measurement of G would not happen until decades after Newton's death.

Question: How did Newton measure the mass of the Earth? (I am assuming it was Newton or was it somebody else?)

Answer: Newton's triumph was to be able to explain Kepler's laws, which empirically describe how the solar system works. You might think he would need the mass of the Sun to do that. But, he could not determine the mass M of the Sun because he did not know the value of the gravitational constant G. What he could determine was their product, GM. Similarly, he could not find the mass of the Earth, only its product with G. For a planet in a circular orbit, $T = 2\pi\sqrt{(R^3/(MG))}$, where T is the length of the year and R is the radius of the circle. A way to get the mass of the Earth is from the acceleration due to gravity $g = 9.8$ m s^{-2}, which can be shown to be $g = M_{\text{Earth}}G/R^2_{\text{Earth}}$, again showing that the product of G with the mass is all you can determine. You can also get GM for the Earth from the period of the Moon (about 28 days). The masses of planets and the Sun were only determined when G was independently measured in the Cavendish experiment, which directly measured the gravitational attraction between two known masses. I believe that G is still the most poorly known of fundamental physical constants. The masses of planets without moons were notoriously hard to determine before we were able to shoot spacecraft at them.

Newton's universal law of gravitation is still used today, particularly for celestial mechanics where trajectories of space probes and satellites are calculated. It has withstood the test of time and qualifies as a truly beautiful theory. However, a more modern theory of gravity, general relativity, has superseded Newton's theory for more recent advances in astrophysics and cosmology.

Question: What is gravity?

Answer: General relativity is the theory that explains this. It is an extremely successful theory and is well-accepted by physicists. General relativity starts with a simple premise, the equivalence principle: there is no experiment you can perform which can distinguish whether you are in a gravitational field or an accelerating frame of reference. For example, if you were in an elevator that was accelerating and a beam of light entered through the side it would follow a curved trajectory to the opposite wall; this is exactly what would happen if you were sitting still in a gravitational field. This principle, coupled with the principle of special relativity (the laws of physics are the same in any inertial frame of reference), leads to the general principle of relativity: the laws of physics are the same in *all* frames of reference. One implication of this theory is that mass deforms spacetime which is how gravity works; mass deforming spacetime is simply a consequence of the postulates of the theory. Is it the last word? Probably not, because gravity has not been reconciled with quantum theory and the quest for a theory of quantum gravity is one of the holy grails of physics.

General relativity is a very mathematical theory in the details, but its implications are pretty easy to understand. The theory finds that spacetime (relativity finds that space and time are not different things) is not Euclidean, but is 'warped' by the presence of mass. An often-used analogy is to imagine a bowling ball placed in the center of a trampoline which causes there to be a sag in the center; now place a marble on the trampoline and, of course, it rolls toward the bowling ball. Be careful to not take this analogy too seriously.

Question: My question is about gravity. In the depictions I have seen of the Einstein model of gravity, planets and stars are shown as depressing a plane of space into a well like depression into which other objects tend to fall. I am OK with this depiction. However, it seems to make the assumption that space is a plane and has only two dimensions. When I observe the Universe, I see three dimensions. It would seem to me that these 'gravity wells' should exist in three dimensions. Why are they only shown as if the fabric of space is like a sheet of paper, in two dimensions rather than in three-dimensional space?
Answer: My stock answer to this kind of question is that the 'trampoline illustration' of deformed spacetime is meant to be a cartoon to illustrate the idea, not an accurate and rigorous representation of the theory of general relativity. You must not take it too seriously or literally. It is also practically impossible to draw a picture of deformed 3D space. To draw deformed 2D space is easy because you use the third dimension to show the deformation; to draw a deformed 3D space would require a fourth spatial dimension which cannot be drawn. And just to make things even more confusing, not just space but also time are components of the space (called spacetime), so a 5D depiction would be required!

General relativity is essentially a single equation, the field equation! Like special relativity, we shall revisit this theory in a later chapter, but its surprising predictions have nearly all been borne out by observations—it is a truly beautiful theory.

2.5 Concluding thoughts on beauty

Does a theory have to be the 'last word' to be beautiful? Absolutely not! In fact, no self-respecting physicist would ever state absolutely that a theory is the last word—any science worth doing should always be looking for ways to move forward. Think of the areas of physics tested for beauty in this chapter.

1. Electromagnetism: this is the closest to being the last word for a theory, but that is not evident from what we have done here. In the mid-20th century the theory was quantized (photons) and later it was combined with another of nature's fundamental forces, the weak interaction; it is now called the electroweak interaction. Very impressive!

2. Relativity: special relativity, still to be discussed in a later chapter, is pretty much the last word for classical mechanics.

3. Gravity: it is sort of in bits and pieces. Newtonian gravity is clearly not the last word, since it has been superseded by general relativity, but it remains useful, crucial even, for many applications. Still, it is a thing of beauty to me. General relativity itself is not the last word because, unlike mechanics and electromagnetism, it has not been quantized; a theory of quantum gravity remains a much sought-after goal for theoretical physics. In my view, it is possible that the current quest to understand dark matter and dark energy may also be indicative of an incompleteness of the theory of gravity. However, both Newtonian gravity and general relativity are beautiful theories in my opinion.

Chapter 3

Physics is surprising

3.1 Introduction

One of the most exciting things about being on the 'front lines' of scientific endeavor is that you sometimes get what you would never have expected. Even more exciting is when you find something which seems to be impossible, which flies in the face of all current wisdom. That's what usually happens when a major breakthrough occurs; science usually proceeds in baby steps, but really important scientific discoveries are giant leaps and are generally followed by a string of unexpected surprises. This chapter focuses on surprises in physics that occurred mainly during the last half of the 19th century and the first half of the 20th century. We have already encountered a few of these in this book, notably the constancy of the speed of light, the distortion of spacetime, and the conceptualization of the notions of force and gravity. For the most part, this chapter will focus on the theories of quantum mechanics, special relativity, and general relativity. My purpose here is not to teach these theories—I did that in the earlier books in the *Ask the Physicist* series, *From Newton to Einstein* and *Atoms and Photons and Quanta, Oh My!* Rather it is to present the consequences of the theories, the unexpected surprises. Be assured that these surprises are not speculations, they have all been thoroughly verified experimentally and are full-fledged physics facts. If anything presented here is a speculation or an unverified prediction, I will be sure to identify it as such.

3.2 Surprises of special relativity

The theory of special relativity, motivated by electromagnetic theory as explained in chapter 2, totally upset all Newtonian ideas of how we think about space and time. Newtonian mechanics, although it works essentially perfectly for everyday life, is incorrect, but this becomes evident only for speeds far faster than any speed we encounter in the everyday world; speeds need to comparable to the speed of light c for relativistic effects to be evident. The fastest man-made object ever launched, the New Horizons mission, which recently acquired data on Pluto and its satellite

doi:10.1088/978-1-6817-4445-2ch3

Charon, has a speed of about $v = 45$ km s^{-1} (100 000 mph) relative to the Sun; the speed of light is about 300 000 km s^{-1}. So $v/c = 1.5 \times 10^{-4}$. The usual test for whether relativity will be important is how large the factor called the gamma factor is compared to 1: $\gamma = 1/\sqrt{(1 - (v/c)^2)}$. For the New Horizon spacecraft, $\gamma \approx 1 + \frac{1}{2}(1.5 \times 10^{-4})^2 \approx 1 + 10^{-8}$, pretty darned close to 1! It is no wonder that Newtonian mechanics worked so well and for so long. But a good physicist wants to know what the laws are even for realms outside our experience. And nowadays there are things, subatomic particles, which get accelerated to very close to the speed of light and are therefore relativistic particles.

The first surprises are referred to as time dilation and length contraction. These are the first nonintuitive predictions of relativity and they are, as the next Q&A will show, not unrelated. Time dilation says that a moving clock runs more slowly than a stationary one and length contraction says that a moving length is shorter than a stationary one.

Question: Lets say it's 25 000 light years to the center of our galaxy. Meaning, light taking off from Earth today would take 25 000 years to get to the center of the Galaxy, or *vice versa*, from the center of the Galaxy to us. Let's say we had a ship that could travel at 99% the speed of light, and we board that ship and leave Earth traveling at 99% light speed. From Earth, as we watched the ship travel, we would see that it took a little bit more than 25 000 years to reach its destination. But what about the perspective of those on the ship? How long would the journey seem to take for the travelers, moving at 99% the speed of light? How much time would have elapsed by their watches?

Answer: The important factor in relativity is $\gamma = 1/\sqrt{[1 - (v/c)^2]}$. In your case, $\gamma = 1/\sqrt{[1 - (0.99)^2]} = 7.1$. Let's call the time the Earth clock measures T; then $T = d/(v/c) = 25\,000/0.99 = 25\,253$ years, a bit longer than the light, as you surmised. The γ factor is the factor by which a clock on the ship runs more slowly than the Earth-bound clock, so $T' = 25\,253/7.1 = 3\,557$ years. But, how can the travelers reconcile this with what they observe? They will agree that they are approaching the center of the Galaxy with a speed of $0.99c$ and that their clock measured 3 557 years when they arrived, so how can they possibly travel such a distance in less than 25 253 years? The reason is simple but surprising—think of the distance from Earth to the destination as a 25 000 ly long stick; they see that stick moving in the opposite direction as they are moving and it is shorter by the γ factor, $d' = d/\gamma = 25\,000/7.1 = 3\,521$ ly. Therefore, they agree that the time of the trip is $T' = d'/(v/c) = 3\,521/0.99 = 3\,557$ years. (Note that v/c is just the velocity in units of light years per year and all distances are in light years.)

The experimental verification for length contraction and time dilation comes from measuring the lifetimes of radioactive particles and also from the excellence of GPS mapping. The following is part of an answer to a question about time dilation on *Ask the Physicist*.

Answer: For example, consider a pi meson, an elementary particle which has a half-life of 18 ns = 18×10^{-9} s. Suppose we make one in the lab with a speed of 80% the speed of light. Since the speed of light is 3×10^8 m s^{-1}, we expect it to go $(0.8)(3 \times 10^8)(18 \times 10^{-9}) = 4.32$ m. But when we actually observe the pi meson, it goes $4.32/\sqrt{(1-0.8^2)} = 7.2$ m. Why? Because this particle is like a clock that ticks once and dies; that clock is moving very fast and therefore ticks much more slowly than if it were just sitting here, so it can go a lot farther than we expect it to. Another good example is the satellites used in GPS systems. Their speeds are quite small compared to light speed, but it is extremely important to measure time very accurately for the systems to be accurate; if corrections for time dilation were not applied, you would never be able to have GPS with near the resolution we all take for granted.

Again, time dilation and length contraction are seen to be two sides of the same coin. I guess you could say one man's time dilation is another man's length contraction. It is important to emphasize that moving clocks really do run slow and moving lengths really do get shorter; some authors will make statements like 'appear to', which is very misleading. How things *look* are often very different from how things *are*, as was shown in book I, *From Newton to Einstein*.

Why these surprising results? The real surprise is more subtle than these examples. What the real difference between the Lorentz and Galilean transformations is that it is no longer assumed, as we had done for millennia, that the rate at which clocks run is independent of where you are or what you are doing—that time is the same for everyone in the Universe. If you look again at appendix C you will see that $t' = t$, implying the assumed universality of time. For completeness, appendix C also shows the Lorentz transformation (which I will not derive nor make any direct use of). In order for the speed of light to be the same for all observers, t' necessarily depends on x and v, just as x' depended on t and v in Galilean relativity, $t' = \gamma[t - (vx/c^2)]$. This is profound because the implication is that space (x) and time (t) are no longer separate entities, but are inextricably entangled, and they are often referred to collectively as *spacetime*.

The next surprising result is actually an offshoot of length contraction—simultaneity. Events which are simultaneous in one frame of reference are not simultaneous in another. This is demonstrated in the next Q&A. (Note also here that length contraction occurs only in the direction of the velocity, not the two perpendicular directions.)

Question: Two frames are in uniform motion along their x-axes. We will consider for simplicity the first frame to be 'fixed', while the second one moves to the right with a velocity v. From the origin of the stationary frame two rays of light are emitted simultaneously, one along the x-axis and the other one at an angle of $60°$ with the x-axis. Two mirrors are placed at the same distance L on the two tracks and the light gets reflected. Obviously, the two reflected rays, as observed in the stationary frame, return to the origin at the same time. I made some calculations and, surprisingly, the two rays, as observed from the moving frame, do not return to the origin at the same time. It is possible that I made a mistake. However, if my calculations are correct, this would be a very strange thing indeed.

Answer: First of all, it would not be 'a very strange thing indeed' to find that what is simultaneous in one frame is not simultaneous in another. One of the keystones of special relativity is that the simultaneity of two events depends on the frame of reference. Figure 3.1 shows the situation as seen by each observer. The primed system moves in the $+x$ direction, while the red-drawn distances to the mirrors are at rest in the unprimed system. The rest lengths of the two distances are L and the angle one makes with the x-axis is θ. The round-trip time along the x-axis is $t_1 = 2L/c$, and along the other length it is the same, $t_2 = 2L/c$. In the moving system all lengths along the x' direction are reduced by a factor $\sqrt{(1 - \beta^2)}$, where $\beta = v/c$; therefore $L' = L\sqrt{(1 - \beta^2)}$, so $t_1' = 2L'/c = 2L\sqrt{(1 - \beta^2)}/c = t_1\sqrt{(1 - \beta^2)}$. For L'' only its x' component is contracted, $L''\cos\theta' = L\sqrt{(1 - \beta^2)}\cos\theta$, while the y' component remains the same, $L''\sin\theta' = L\sin\theta$; from these you can easily show that $L'' = L\sqrt{(1 - \beta^2\cos^2\theta)}$ and $\tan\theta' = (\tan\theta)/\sqrt{(1 - \beta^2)}$. Finally, $t_2' = 2L''/c = 2t_2\sqrt{(1 - \beta^2\cos^2\theta)} \neq t_1'$—the two light pulses do not arrive at the origin simultaneously in the moving frame.

We have already been surprised by the constancy of the speed of light regardless of the motion of the source or observer. This will surely suggest that all velocities may behave surprisingly. Suppose that you are in a car (A) driving down the road with a velocity of $v_{AR} = 60$ mph. Another car (B) is driving on the same road with a speed of 40 mph but in the opposite direction, so its velocity is $v_{BR} = -40$ mph. With what velocity does car B see you approaching it? Easy, right? Obviously 100 mph. However, even though this is obvious, it is helpful for such calculations to have a general equation which is called the velocity addition formula: $v_{AB} = v_{AR} + v_{RB}$. In all this discussion you should read v_{IJ} as 'the velocity of I with respect to J'. To apply this to our simple little problem, first note that $v_{RB} = -v_{BR} = +40$ mph. So, finally, $v_{AB} = 60 + 40 = 100$ mph. However, this is wrong, as intuitively true as it may be to you. The relativistically correct velocity addition formula is $v_{AB} = (v_{AR} + v_{RB})/[1 + (v_{AR}v_{RB}/c^2)]$. Because the speeds are so small, the denominator is, for all intents and purposes, equal to 1. As the next question demonstrates, if speeds are not small compared to c, your intuition will fail you. (For this question, keep in mind that $c = 300\ 000$ km s^{-1}.)

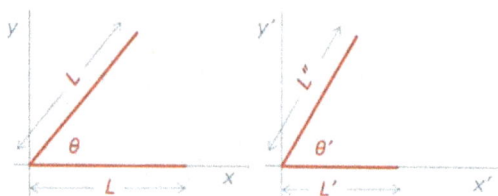

Figure 3.1. The 'tracks' in the stationary (x,y) and moving (x',y') systems.

Question: OK, I am having trouble with this speed of light thing. If a train left Earth at near the speed of light and turned on its light, the light would leave the train at the speed of light relative to the train, correct? But doesn't that mean that those photons are traveling away from Earth at *way* more than the speed of light? If so, could we launch a ship (hypothetically) and get to 90% the speed of light and then launch another smaller ship from that ship and travel 90% of light speed relative to that ship? A ship that would be distancing itself from Earth at *way* more than the speed of light? This is where our slow speeds in space bother me. Since space is a vacuum (nearly) and rockets only push on themselves from the inside and speed is relative, wouldn't a ship that is coasting 40 000 mph away from the Earth suddenly fire a rocket and accelerate no matter what? Without wind resistance and with the slight help of solar winds, I would think we could travel a probe much quicker. We might have to build it in space, but 8 Saturn Vs connected together ought to make a difference. No?

Answer: The problem you are having is that you are thinking in terms or classical physics or intuitively; intuition is based on experience and you have no experience with trains going nearly the speed of light. Whenever you extrapolate what you 'know' to be true of trains and light to situations outside your experience, you need to be cautious. You assume that if the train has a speed v relative to the Earth and the light has a speed c relative to the train, then surely the speed of that light relative to the Earth will be $c + v$. This is called Galilean velocity addition and is only approximately true when all speeds are very small compared to c. In the theory of relativity the velocity addition expression is $u' = (v + u)/[1 + (vu/c^2)]$, where v is the speed of one thing relative to the Earth, u is the speed of another thing relative to the first thing, and u' is the speed of the other thing relative to the Earth. In your first case, v is the speed of your train and u is c, the speed of light, so $c' = (v + c)/(1 + v/c) = c$. In your second case, $v' = (0.9c + 0.9c)/(1 + 0.81) = 0.99c$.

The constancy of the speed of light is also built into velocity addition. Suppose that there is a light beam whose speed is measured as $v = c$ when you are at rest and then you move with speed $u = \frac{1}{2}c$ toward the source. With what speed v' do you see the light approaching you? Then $v' = (\frac{1}{2}c + c)/(1 + (\frac{1}{2}c^2/c^2)) = c$. Surprising, indeed! In fact, it is mathematically impossible to get a value of v' greater than c—you cannot use a Lorentz transformation to get to a speed greater than c.

Of course, the most famous surprise from the theory of special relativity was the equivalence of mass and energy, $E = mc^2$. A small amount of mass can be converted into a vast amount of energy.

Question: I know Einstein said $E = mc^2$ and basically all matter can be equated to some quantity of energy; then why do we go to the gas station to fill our cars? Why can't we use garbage, which is mass and has energy, to power our cars? How can we convert matter to energy? I know we can burn gasoline to use perhaps ¼ the heat content in the form of expanding gas to apply pressure to the piston in the engine. Has anyone invented a matter converter that changes matter to energy yet?

Answer: Most of the energy mankind uses comes from chemistry. Burn coal or gasoline, for example. When you eat and metabolize food, chemistry is going on. The energy which is extracted comes from—guess what—mass! For example, when you burn coal the main thing which is happening is that carbon is combining with oxygen to form carbon dioxide. One carbon dioxide molecule has a smaller mass than one carbon atom plus one oxygen molecule. So, chemistry is the best-known example of your 'matter converter'. The problem is that an extremely tiny fraction of the mass is converted to energy, something like 0.00000001%, so chemistry is a very inefficient source of energy. Now, to get more efficiency we have work not with atoms but with the nuclei of atoms. If a heavy atomic nucleus can be induced to split (fission), the mass of the fragments is smaller than the mass of the initial nucleus by an amount that is much bigger than with chemistry, something like 0.1%, which is a huge improvement over chemistry; this is how nuclear reactors and bombs work. Also from nuclear physics, you can take very light nuclei and make them combine (fusion) and get something like 1% of the mass converted into energy; this is how stars work and so, you see, solar energy comes from 'matter converters' too and so does wind energy since the Sun is the energy which causes winds to blow. If you want to get 100% efficiency you have to go to particle–antiparticle interactions in particle physics. When an electron and its antiparticle the positron meet, their mass completely disappears and all the energy comes out as photons. Did you ever see the *Back to the Future* movies? Doc came back from the future with a small appliance called 'Mr Fusion' which had been invented to do precisely what you want to do—convert garbage into the huge amount of energy needed to power the time machine.

Finally, we should mention the universal speed limit, the surprising outcome of the theory which forbids any object from exceeding the speed of light. We have already shown that you cannot exceed c by using the velocity addition. But it seems nonintuitive that, if we push hard enough, add enough energy, we could not eventually be able to reach any speed. Actually, the equation $E = mc^2$ is only the correct form of this equation if the object is at rest; if the object is moving it has more energy, which is not surprising since an object in motion has kinetic energy. You might recall that kinetic energy in Newtonian mechanics is $\frac{1}{2}mv^2$, so you might guess that the correct form for E would be $E = mc^2 + \frac{1}{2}mv^2$, but you would be wrong! Don't forget, Newtonian mechanics is no longer a correct theory, it is only approximately correct for low velocities. The correct form is $E = \gamma mc^2$ (remember γ?); it is fairly easy to show that, at low velocities, $\gamma mc^2 \approx mc^2 + \frac{1}{2}mv^2$, which is shown in appendix F.

Question: I know that nothing can travel at or faster than the speed of light. But, just simply, why? What equations or whatever say no?

Answer: The energy of a particle is $E = \gamma mc^2 = mc^2/\sqrt{(1 - (v^2/c^2))}$, so the energy required to accelerate the mass to the speed of light is infinite and there is not an infinite amount of energy in the Universe. Another way to look at is that the equation for energy could be interpreted as $E = Mc^2$, where $M = \gamma m$ and M is the inertial mass of a particle moving with speed v and whose mass is m if at rest; in other words, the mass of an object increases as it increases its speed. Therefore it gets harder and harder to accelerate it as it goes faster and faster. Note that as v approaches c, M approaches ∞, so it is impossible to push beyond c.

I have often implied in this section that results are surprising because they are nonintuitive. This is true. But what is intuition based on? In fact, intuition, what we expect, is based on our experience and we have never experienced speeds so large and therefore intuition may quite possibly be wrong if we try to apply it outside our experience. It is dangerous to extrapolate your intuition beyond the limits of your experience.

3.3 Surprises of general relativity

General relativity extends the theory of special relativity to include frames that are accelerating. Added now to the principle of relativity is the equivalence principle, which states there is no experiment which you can do to distinguish whether you are in a uniform gravitational field with gravitational acceleration g or you are in empty space in a frame which has acceleration g relative to some inertial frame. This leads to our first general relativity surprise.

Question: What causes gravity? How can gravity be explained? General relativity, as I understand it, says that gravity is not a force or interaction. Rather that spacetime is 'curved' by the presence of mass, and that this curve 'tells' other matter how to behave. Have I got that right? But the question remains, does it not? Accepting what general relativity says is one thing, but in reality the real question is why or how does mass cause spacetime curvature? Am I thinking correctly here? I teach astronomy at a local school, and some of those kids come up with some tough (for me) questions.
Answer: General relativity starts with a simple premise, the equivalence principle: there is no experiment you can perform which can distinguish whether you are in a gravitational field or in an accelerating frame of reference. For example, if you were in an elevator which was accelerating and a beam of light entered through the side it would follow a curved trajectory to the opposite wall; this is exactly what would happen if you were sitting still in a gravitational field. This principle, coupled with the principle of special relativity (the laws of physics are the same in any inertial frame of reference), leads to the general principle of relativity: the laws of physics are the same in any frame of reference. One implication of this theory is that mass deforms spacetime, which is, as you

state, how gravity works; mass deforming spacetime is simply a consequence of the postulates of the theory. Is it the last word? Probably not, because gravity has not been reconciled with quantum theory and the quest for a theory of quantum gravity is one of the holy grails of physics. I would not say that gravity is not a force just because we understand the mechanism for that force. Asking 'why or how' mass causes the curvature is essentially equivalent to asking what is mass, why do objects possess it? The answer to the why or how question is simply that general relativity predicts as an inevitable consequence that the presence of mass will cause spacetime to be warped. In other words, the attractive force between masses is explained by the behavior of the four-dimensional space we reside in. That is much more satisfying, I believe, than Newton's universal gravitation, which simply says that force exists because it is.

In the elevator example just cited, the amount of fall would probably be too tiny to measure, because light goes so fast that it will not have much time to fall as it crosses the elevator. But imagine that instead we look at light passing a star, which will have a much stronger gravitational field. For example, at the surface of the Sun, $g = 274$ m s^{-2}, about 28 times stronger than on Earth. The first test of general relativity was done during a total solar eclipse in 1919 when Arthur Eddington observed a star, the position of which was well known, to be in slightly the wrong position because of the deflection of the light by the massive Sun. The deflection was just as had been predicted by Einstein and overnight he was an international 'rock star'! (Disclosure: the use of the equivalence principle alone to argue why light should be deflected is very useful to convey an understanding of why light is deflected by gravity. Because of a mathematical subtlety of general relativity, however, it is incorrect. The actual deflection is only half that you would calculate from the acceleration g using the elevator example.)

Over many years more examples of deflection of light by large masses have been observed. The next question reveals one of the most impressive, Einstein's cross.

Question: Can gravity focus light like a lens?
Answer: Gravitational lensing is light being focused by gravity. An example of gravitational lensing is shown in figure 3.2: 'Einstein's cross', which shows four images of the same quasar. Doing the imaging here is a massive galaxy located between us and the quasar.

Closely related to deflection of light by gravity is the gravitational red shift. If gravity has an effect on light moving perpendicular to the field, as just demonstrated above, it must also have an effect on light moving parallel to the field. For a baseball thrown horizontally, the effect is to curve downward. For a baseball thrown vertically, the effect is to slow down. Light behaves the same for the first effect, but the second effect is not possible because the light must move with constant velocity. So, perhaps the second effect should be restated: for a baseball thrown vertically, the effect is to lose energy. Light leaving a star, for example, will lose energy as it gets farther away

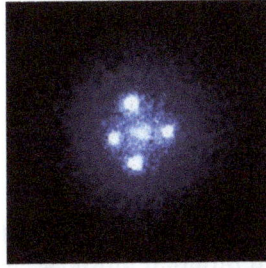

Figure 3.2. Einstein's cross.

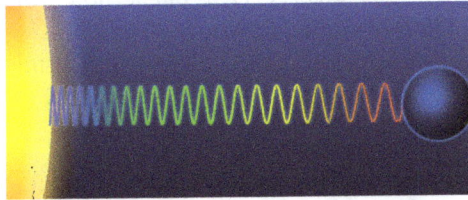

Figure 3.3. Cartoon of gravitational red shift.

(figure 3.3); when light loses energy (more about this in section 3.4), its frequency gets smaller and so its wavelength gets longer. So the wavelength shifts slightly toward the red; the effect is therefore called the gravitational red shift. This effect is quite small, but has been measured in astronomical observations; this is difficult, because it is usually quite small compared to conventional red shift (Doppler effect) caused by receding velocities. It has also been observed in Earth-based experiments over a distance of just 22 m by extraordinarily accurate frequency measurements. Of course, if light is approaching a large mass it gains energy—a gravitational blue shift.

Closely related to the gravitational red shift is gravitational time dilation. It turns out that, because of the red shift, the closer a clock comes to a source of gravity, the more slowly it runs. So a clock on the surface of the Earth will run more slowly than one that is very far away.

Question: If I am at a point in space where very, very, very little gravitational pull exists, what happens to the time clock on my spaceship as opposed to clocks on Earth? What I am asking is, if you were at a point in space where the gravitational forces were almost 0, would the clock in your spacecraft be running extremely fast as observed from Earth?

Answer: Choose some place, maybe halfway between here and the Andromeda galaxy, our nearest galactic neighbor. For all intents and purposes the field there is zero and a clock would run at some rate. Now, take that clock and put it on the surface of a sphere of radius R and mass M. The gravitational time dilation formula you need to compute how much slower the clock would tick at its new location is $\sqrt{[1 - (2MG/(Rc^2))]}$, where $G = 6.67 \times 10^{-11}$ N · m^2 kg^{-2} is the universal constant of gravitation and $c = 3 \times 10^8$ m s^{-1} is the speed of light. Putting in the mass and radius for the Earth, I find that the space clock runs about

$7 \times 10^{-8}\%$ faster than the Earth clock, certainly not 'extremely fast'! The bottom line here is that, except very close to a black hole, gravitational time dilation is a very small effect. An interesting fact, though, is that corrections for this effect must made in GPS devices because they require extremely accurate time measurements to get accurate distance measurements.

Gravitational time dilation has been fully verified by experiments such as comparing atomic clock rates at different altitudes. And, of course, any experiment which verifies light deflection also verifies time dilation.

Mentioned in the previous answer is a black hole. Black holes are also a prediction of general relativity and, again, a surprise. The theory predicts that under the right conditions a large mass can collapse into a singularity, a single point, which is called a black hole for reasons we will see below.

Question: What causes the birth of a black hole?
Answer: When a star has exhausted most of its hydrogen fuel, it begins to collapse under its own gravity. The next thing that happens is that it explodes—a supernova. Then, providing that the star was massive enough, it will continue collapsing until it has essentially collapsed into zero volume, a black hole.

As you can imagine, the gravitational field can be huge near a black hole because its mass is large and you can get very close. To get a perspective on this, consider relativistic time dilation. From the previous answer, $t_r = t_\infty \sqrt{[1-(2MG/(rc^2))]}$, where t_∞ is the time rate with zero field, far away from the black hole, and t_r is the time rate at r from the black hole. Note that when $r = r_S = c^2/(2MG)$, $t_r = 0$—time stops! This is called the Schwarzschild radius and is also referred to as the event horizon. Nothing, not even light, can escape from inside this radius; this is the origin of the name black hole, because no light at all escapes from this region in space, which would appear to be a perfect black sphere. To get a feeling for the size of a 'small' black hole, $r_S \approx 3$ km for a black hole of three solar masses.

Question: I was on a training course and learned that the Andromeda galaxy revolves around a black hole. Is that true of all galaxies?
Answer: Astrophysicists believe that nearly all large galaxies have a supermassive black hole at their center.

This is a fairly recent finding. Supermassive means from hundreds of thousands up to billions of solar masses!

It would seem, since nothing can escape a black hole, that a black hole is forever and, depending on conditions around it, will either keep growing as it vacuums up the stars and other stuff around it, or will keep the same mass, never losing mass. However, that is not the case.

Question: In the event horizon of the black hole, a pair production of a particle and an antiparticle and the falling of one into the black hole and the releasing of the other from it gives us Hawking radiation, but my question is, isn't it possible for both the particle and the antiparticle to fall into the black hole. If that happens, the energy of the Universe decreases, so how can the Universe compensate this decrease in energy to validate the law of conservation of energy.
Answer: There are two ways you can imagine the creation of the particle–antiparticle pair.

1. Vacuum fluctuations, where a particle–antiparticle pair is spontaneously created just outside the event horizon, violating energy conservation. However, the uncertainty principle allows this violation, but only for a very short time. If one escapes, it adds energy to the Universe, so the captured particle must have negative total energy, thereby decreasing the mass of the black hole and maintaining energy conservation. If both particles are captured, they are required by the uncertainty principle to recombine, thereby keeping the mass of the black hole constant.

2. When a virtual pair is created just outside the event horizon, the intense gravitational field of the black hole can provide the energy for the PAP to become real rather than virtual. In this case, the energy acquired by the pair will be lost by the black hole, reducing its mass. If only one of these is captured, the energy of the escaped particle will be added to the Universe, but the energy of the captured particle will be added to the (already reduced) energy of the black hole, resulting in a net loss of energy of the black hole equal to the energy of the escaped particle.

The energy of the Universe is conserved in both scenarios.

Finally, one of the last tests of general relativity has very recently been completed. Since spacetime is viewed as 'warpable', we might think of it as a flexible sheet which could support traveling waves.

Question (submitted by my daughter!): Is the observation of gravitational waves exciting for you? I don't really understand it but I'm trying to—it seems cool!
Answer: Observing gravitational waves has been a holy grail of physics since before I was an undergraduate (like 55 years ago, gasp!). So yes, it is pretty exciting news. This is actually just the first *direct* evidence for gravitational waves. Indirect evidence was found when a pair of stars were observed orbiting each other and spiraling in toward a collision. The loss of energy turned out to be exactly equal to the amount of energy they would lose if radiating gravitational waves. A Nobel prize in physics was awarded in 1993 for this observation. (Incidentally, it is *gravitational* wave, not *gravity* wave. Gravity waves are a phenomenon in fluid dynamics involving the interface between two media like the air and the ocean; waves on the surface of the ocean are gravity waves.)

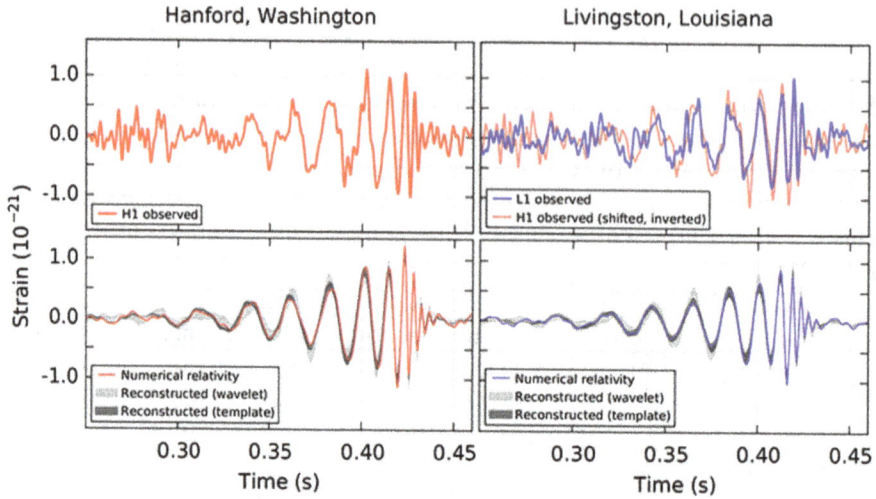

Figure 3.4. Gravitational wave data (above) and theoretical calculations (below).

The recent (15 September 2015) observation by the LIGO collaboration of gravitational waves represented the culmination of an almost century-long quest to observe the waves in spacetime predicted by general relativity; it was one of the last predictions of general relativity to have been verified. Two extraordinarily sensitive detectors, in Hanford, Washington and Livingston, Louisiana, simultaneously observed the gravitational waves. Figure 3.4 shows the excellent fits to the data by general relativity calculations. The calculations identified the event to result from two black holes spiraling in and merging.

Question: If all things with mass have a proportional amount of gravity and gravitational waves were recently observed to have been produced by two black holes converging, is it correct to deduce that all things with mass produce gravitational waves proportional to their mass?

Answer: Yes. Any object with mass which accelerates should radiate gravitational waves. Wave to someone and you cause gravitational waves. However, gravity is the weakest force in nature, so, for the waves to have a big enough amplitude to be detectable, the masses must be extremely large (as in black holes).

Question: How fast do gravitational waves move? Is that rate constant?

Answer: The speed of gravitational waves has never been measured. The speed at which a gravitational field propagates should be the same, but it has never been measured either; this would determine, for example, how long it would take until the Earth stopped orbiting if the Sun suddenly disappeared. The theory of general relativity, which predicts gravitational waves, says that the speed of gravity should be the same as the speed of light. The recent observation of gravitational waves determined the distance to the source to be about 1.33 billion light years, but the uncertainty was very large, about 40%, so it really provides no measurement of the speed. Whatever the speed, there is no reason to think that it would not be constant everywhere in empty space.

Question: Since the Earth is trapped in orbit around the Sun by the Sun's gravity, how can a wave moving out from an object in turn pull a second object back towards the object that generated the wave? In other words, how can gravity waves moving out from the Sun at the same time pull the Earth towards the Sun?
Answer: You are confusing gravity with gravitational waves. Our picture of gravity is that mass (like the Sun) deforms the space around it. Something like the Earth orbiting the Sun is not really feeling a force, it is following the contour of the space. But, if something accelerates, like the Earth going around in its orbit, it will send out ripples, and those are the gravitational waves. The waves the orbiting Earth is sending out are far too tiny to be seen. But, if the objects orbiting are much more massive, like the two orbiting black holes which were observed in the recently reported observation of the waves, the waves are much bigger.

The detectors both have two 4 km long arms. A third is under construction in India. When all three events are timed, the direction from which the waves come will be determined more accurately. The use of gravitational waves as a new way to observe the Universe has just begun (figure 3.5)

The final surprise in general relativity that I want to discuss are the discoveries in cosmology in recent years of the phenomena called *dark matter* and *dark energy*. I only label these as general relativity surprises because they are discoveries related to gravity but not predicted by general relativity.

Question: Can you please tell me what dark matter is.
Answer: There are many instances where astronomical objects behave as if there were more mass present than we can see. The best-known example is the

Figure 3.5. The future of gravitational wave research?

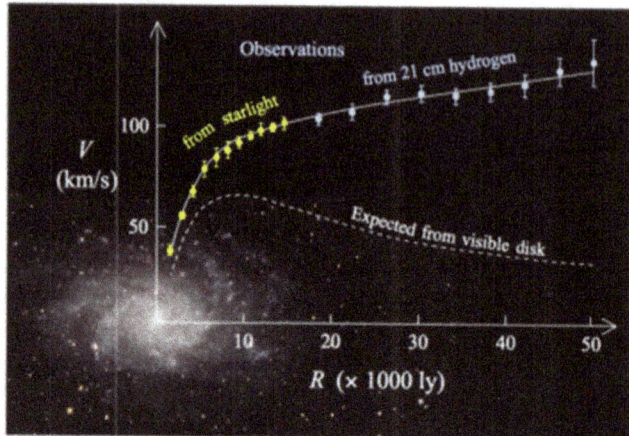

Figure 3.6. Rotational velocities of stars in a galaxy as a function of their distance from the center.

dynamics of galaxies where the speeds of stars far from the centers of the galaxies are much bigger than expected on the basis of the mass that can be seen; see figure 3.6. The conclusion usually drawn is that there is much more mass present than we can see and this hypothesized stuff is called *dark matter*. Nobody knows what it is and nor has it yet been detected directly.

Question: I've been reading about dark matter and dark energy lately, and like most people, I'm very confused! I understand that dark energy is the energy of empty space, representing the cosmological constant.

Answer: The motion of many things in the Universe cannot be understood if we apply known laws (gravitation) using observable mass (stars, planets, etc). Figure 3.6 provides one example. Therefore, it is postulated that there must be something, dubbed *dark matter*, that we cannot see that is causing things to move differently than we expect. All attempts to observe this stuff have failed, although the search goes on. My own take on dark matter is that it is a possibility that our 'known laws' are not as good as we believe them to be; that is, we do not understand gravity as well as we think we do. Several years ago it was determined that the most distant objects are not just moving away from us very fast, they are actually accelerating. This was totally unexpected because gravity is, as we know it, purely an attractive force and the speeding up would imply the existence of a repulsive force. The origin of this mysterious force is referred to as *dark energy*. Again, what it is is not well understood and one way to integrate it into general relativity (the theory of gravity) is, as you note, to reintroduce the cosmological constant rejected by Einstein early in the development of the theory.

Question: What is the cosmological constant? Was it only because of it that Einstein thought the Universe was static?

Answer: When Einstein proposed the theory of general relativity around 1918, it was generally believed that the Universe was static. General relativity is the theory of gravity, and if gravity is the only interaction among stars and galaxies, this is not possible; with only gravity, the Universe would have to be either expanding and slowing down or collapsing and speeding up. Therefore Einstein had to introduce something to balance the universal gravitational force; this something was called the cosmological constant. It was later discovered that the Universe is expanding and he later denounced the cosmological constant as 'my greatest blunder'. Interestingly, several years ago it was discovered that the expansion of the Universe is actually speeding up, which implies some kind of repulsive force, akin to the cosmological constant. The cause of this acceleration is referred to as dark energy; this has reignited interest in Einstein's supposed 'greatest blunder'.

Here are two instances of surprises which fly 'in the face of all current wisdom' in the words of the introduction to this chapter. It makes for exciting times when well-established phenomena are not well understood.

3.4 Surprises of quantum mechanics

Quantum mechanics, like special relativity, had its origins in classical electrodynamics. As emphasized in chapter 2, it was well-established by the beginning of the 19th century that light is a wave; by the end of the 19th century, it was well-established that the waves were electromagnetic. The wave-like nature of light supplanted Newton's conclusion that light was particles. The results of two experiments defied theoretical explanation using electromagnetic waves.

Question: Light is considered as a particle because of the photoelectric effect. Are there any other experiments that show light is made up of particles?
Answer: Historically, there is an equally important experiment demonstrating that light *sometimes* behaves like a particle—Compton scattering. If electromagnetic radiation (x-rays were used in the original experiment) falls on a solid, the scattering from electrons in the solid can only be explained with particles, not waves. In addition, any time atoms or nuclei de-excite, they emit single photons.

Compton scattering and the photoelectric effect are described in more detail in the second book of the *Ask the Physicist* series, *Atoms and Photons and Quanta, Oh My!* In the photoelectric effect, light falling on a metal plate causes electrons to be ejected from the plate; in Compton scattering, light directed onto a solid is observed to scatter off the electrons in the solid. The photoelectric effect was shown in 1905 by Einstein to be explicable only if light of frequency f was, in fact, a stream of massless particles each with energy hf, where $h = 6.63 \times 10^{-34}$ m^2 kg s^{-1} is Planck's constant. This was a truly surprising hypothesis, because, if correct, how could all the experiments explained by waves be understood? A radical view of this conundrum emerged—wave-particle duality.

Question: Is light a particle or a wave? Which is right and why?
Answer: Light is not a particle *or* a wave, it is a particle *and* a wave. This is called wave–particle duality. If you design an experiment to observe a particle, you will observe a particle; and, if you design an experiment to observe a wave, you will observe a wave.

In 1923, Arthur Holly Compton showed, both experimentally and theoretically, that light, when scattered from electrons, behaves precisely like massless particles of energy $E = hf$ scattering from electrons. In 1924 the French physicist Louis deBroglie proposed that if waves were also particles, then particles must also be waves. deBroglie's hypothesis was confirmed in 1927 by Davisson and Germer of Bell Laboratories, who scattered electrons from a nickel crystal and observed diffraction, which was only attributable to wave properties.

Question: What do you think about particle–wave theory? And does the particle–wave theory negate Heisenberg's uncertainty principle?
Answer: What do I think about it? I think that wave/particle duality is a fact of nature. Why would you think that it would 'negate' the uncertainty principle? You could actually say that duality results from the uncertainty principle, or *vice versa*.

Back when Newtonian mechanics reigned supreme in physics, a statement was often made that if you know the laws of physics and are able to know exactly the position and velocity of every object in the Universe, you can predict with certainty everything that will ever happen and can reconstruct everything that has happened. Even though this could never be actually done, given the size of the Universe, the complexity of the laws of physics, and the paucity of computing abilities, it still raises serious philosophical issues—it implies a fatalistic picture of the Universe because everything is preordained. The notion of free will goes out the window. Surprise! Wave–particle duality essentially implies that knowing all the positions and velocities is not just difficult, it is absolutely impossible. The Heisenberg uncertainty principle states that there is a limit to how accurately you can simultaneously know both the position (x) and linear momentum ($p = mv$), $\Delta x \Delta p \approx \hbar$, where $\hbar = h/(2\pi)$. The more accurately you try to measure the position, the less accurately you can know the momentum, and *vice versa*. If you want to determine the momentum (velocity, essentially) to perfect accuracy, $\Delta p = 0$, you lose all knowledge of the position, $\Delta x \approx \hbar/0 = \infty$. Momentum and position are called conjugate variables. There are other conjugate variable pairs, for example energy and time: $\Delta E \Delta t \approx \hbar$. To know the energy of a system accurately, you must observe it for a very long time. Another way to look at the uncertainty principle is through something you often read in physics popularizations or attempts to transfer physics ideas to literature and the humanities: measuring one variable disturbs the value of another variable. (To understand the connections between duality and the uncertainty principle better, see chapter 2 of *Atoms and Photons and Quanta, Oh My!*.)

Another big surprise of quantum mechanics is the behavior of very small systems, atoms, molecules, and nuclei. The only way the properties, structures, and interactions of these things could be understood was by quantizing various variables, notably energy and angular momentum. This first appeared in Niels Bohr's model of the hydrogen atom. He proposed that the angular momentum of the electron in a hydrogen atom could only have certain values; these quantized values then led to quantized energy levels of the atom. If the atom was in an excited state, it would drop to a lower state by emitting a photon whose energy was just the energy difference between the two states. This was a startling surprise. In classical systems, there is no restriction on how much energy or angular momentum you can add. A wheel spinning with angular velocity ω and of mass M and radius R has an energy $\frac{1}{4}MR^2\omega^2$ and angular momentum $\frac{1}{2}MR^2\omega$; you may speed up the wheel by any amount you want, thereby adding any amount you want to the energy and angular momentum. But if you have a very tiny system like a diatomic molecule that is rotating about its center of mass, you cannot just add any amount of energy to it; it may be possible for it to have angular velocity ω or $\omega + \varepsilon$, but not $\omega + \frac{1}{2}\varepsilon$.

Chapter 4

Physics is cool

4.1 Introduction

One could say that I am a 'physics junkie'. Sometimes I get questions which really grab my attention. I will sometimes spend days solving the problem, or polishing it, or looking at it from different perspectives, or just plain trying to figure it out. Here I collect a few of my favorite Q&As for some of the 'coolest' problems I have dealt with. Be warned that some of these answers are long and can get a little technical. But, like all of this book, you can skip over something that is too technical or too boring to you. Part of the reason the answers are long, though, is to get enough verbiage to make the mathematical parts as clear as possible, so you might at least give them a try!

4.2 Cool stuff

This is not really a hard problem, but it does require a little facility with Newton's second law—both the translational and rotational forms thereof. It is one of those problems where you look at it and think 'I should be able to just solve this by inspection, just intuitively know what the answer should be.' Imagine a wheel rolling along with some velocity and rolling onto a conveyor belt which has the same velocity. After a while, how fast is the tire rolling?

> **Question:** A tire rolling on a level surface at a linear speed of 10 mph rolls on to a conveyor belt which is also moving at 10 mph in the same direction. How will the tire's speed change? Will it be 20 mph? Or 10 mph? Will its rotation stop? Or reverse?
>
> **Answer:** Figure 4.1(a) shows the situation when the rolling tire is about to touch the conveyor belt. It is rolling without sliding, so the point of contact with the floor is at rest, the center is moving forward with a speed v (your 10 mph), and the top is moving forward with speed $2v$; the belt is also moving forward with speed v. I find this problem much easier to do if I transform into a coordinate system

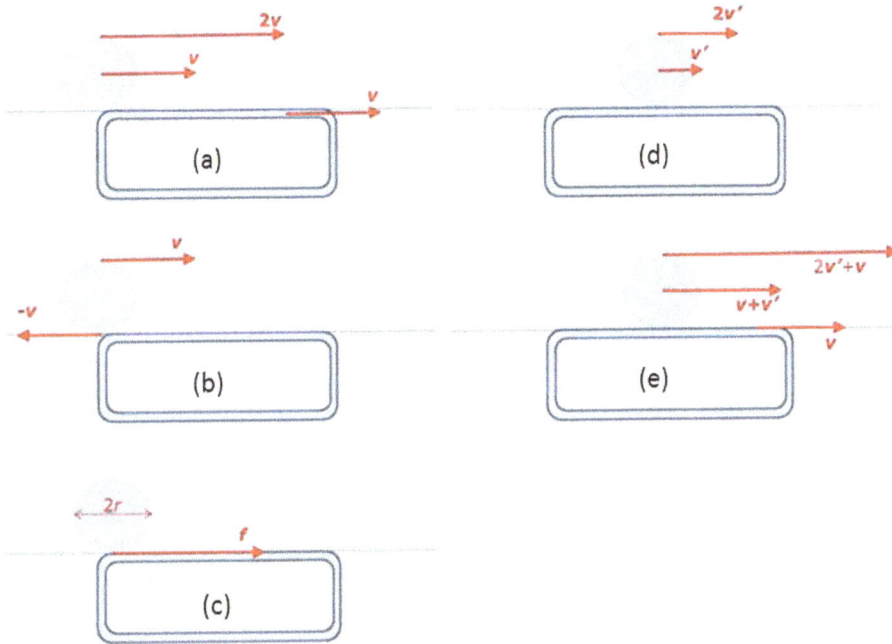

Figure 4.1. Wheel rolling onto a conveyor belt: (a) just before contact, (b) transform into wheel frame, (c) contact, (d) after sliding stops, and (e) transform back.

which is moving with speed v to the right; in that coordinate system the belt (upper surface) and tire are both at rest and the top and bottom edges of the tire have speeds v, as shown in figure 4.1(b). Before getting into the hard part, there is a special case which we can get out of the way first: if there is no friction, the tire will continue on its merry way unchanged, with both its speed and its angular velocity unchanged, because there are no net forces or torques on it.

Now, as soon as it gets on the belt there will be a frictional force f trying to accelerate it to the right, as shown in figure 4.1(c), so it will start sliding along the belt. The frictional force will be $f = \mu mg$, where μ is the coefficient of kinetic friction, m is the mass of the tire, and g is the acceleration due to gravity. Call v' the speed which the center acquires in some time t. Then Newton's second law for translational motion is $m\Delta v = \mu mgt = mv'$, so $v' = \mu gt$.

There will also be a torque $\tau = fr = \mu mgr$ that acts opposite the direction the tire is rotating. There will come a time when the bottom edge of the tire will be at rest relative to the belt because of this torque; this is shown in figure 4.1(d). Newton's second law for rotational motion is $\Delta L = \tau t = -(\mu mgr)(v'/\mu g) = -mv'r$, where L is the angular momentum of the system. The angular momentum is the angular momentum about the center of mass plus the angular momentum of the center of mass. $L_1 = I\omega_1 + 0 = Iv/r$, where I is the moment of inertia about the center of mass; $L_2 = I\omega_2 + mrv' = (v'/r)(I + mr^2)$. Therefore, $(v'/r)(I + mr^2) - Iv/r = -mv'r$, or, $v' = Iv/(I + 2mr^2)$. This is surprisingly simple, and it is particularly surprising that it does not depend on μ. However, the time to stop sliding does depend on μ,

$t = v'/\mu g$; the slipperier the surface, the longer it takes to stop slipping, which is not unexpected.

Finally, we need to transform back into the original coordinate system by simply adding v, as shown in figure 4.1(e); the new speed forward is $v + v'$ and the new angular velocity is v'/r. As an example, suppose we model the tire as a uniform cylinder of mass m and moment of inertial $I = \frac{1}{2}mr^2$; then $v' = v/5$, or in your case, 2 mph, so the tire is rolling without slipping with a speed of 12 mph.

I have entertained many questions about artificial gravity in rotating space stations over the years. Here is one of them.

Question: Imagine a cylinder in space, rotating at the appropriate velocity so that objects resting on the rounded surface of the interior experience a centrifugal force as if they were experiencing normal Earth gravity. Assuming that there is an atmosphere inside this cylinder, a hot-air balloon with enough lift to rise in the atmosphere will travel 'upwards' relative to the surface; how would the balloon behave in the rotating 'space station'? Would the observer piloting the balloon find himself rising at a vertical angle relative to the surface, rather than straight upwards? Would he eventually collide with the surface at another point? In short, what would the path of the rising balloon be in the centrifuge?

Answer: I love questions about 'artificial gravity' in rotating space stations; there are sometimes surprising results unless you are standing still on the interior surface and the radius is very large compared to your height. The most important thing to realize is that this is not really like gravity at all, because the instant you lose contact with the surface you experience no forces at all so you move in a straight line; nevertheless, with a large enough radius, the behavior of the path as viewed by an observer on the surface can be quite similar to how it would be on Earth, particularly for the straight-up jump. We will also need to know that for the centripetal acceleration of a cylinder of radius r_0 to be $g = 9.8$ m s^{-2}, the angular velocity ω will be $\omega = \sqrt{(g/r_0)}$.

Now, why does a hot-air balloon rise? It is because of the buoyant force, and the buoyant force arises because the pressure on the bottom of the balloon is larger than the pressure on the top. So the first question we need to answer is whether there is a pressure gradient in the space station (i.e. pressure decreases with distance from the surface) and how that compares with the pressure gradient in a gravitational field. In the space station there is certainly a gradient because (as you note) it is simply a giant centrifuge, so air will tend to move to larger radii. For an incompressible fluid there is a centrifuge equation for pressure a distance r from the center, $P(r) = P_0 - \frac{1}{2}\rho\omega^2(r_0^2 - r^2) = P_0 - \frac{1}{2}(\rho g/r_0)(r_0^2 - r^2)$; here r_0 is the distance from the center to the outer surface, $P_0 = P(r_0)$, and ρ is the fluid density. Technically, this is not correct because the density varies with r, but we are really only interested in situations where the density is fairly constant; for example, if $r_0 = 1000$ m, $r = 900$ m, and $P_0 = 10^5$ N m^{-2} (approximately atmospheric), $P(r) = 0.99 \times 10^5$ N m^{-2}. In a uniform gravitational field there is an empirical equation to calculate pressure as a function of altitude h, $P(h) = P_0$ $[1 - (Lh/T_0)]^{gM/(RL)}$. Without going into detail, I find that $P(h = 100$ m$) = 0.98 \times 10^5$ N m^{-2}; we can therefore conclude that the buoyant force in the space station will be very similar to that on Earth, at least for the first few hundred meters.

I want to look at what the balloon does from two perspectives—an observer outside the space station and the guy inside who releases the balloon.

1. From outside (figure 4.2(a)), the instant that the balloon is released it experiences no force except the buoyant force B. At that instant it has a tangential speed of $v = r_0\omega = \sqrt{(gr_0)}$. The buoyant force is constant in magnitude but changes in direction since it always points toward the center of the space station. So as the balloon is carried in the direction of the initial tangential velocity and rises due to B, the direction of both B and v changes and the magnitude of v changes. If the radius is very large (obviously not the case in figure 4.2(a)), the direction of B will not change much and the path followed will approximately be a parabola.

2. When looking from inside the rotating cylinder, there are two fictitious forces, the centrifugal force $F_r = mg$ and the Coriolis force $F_c = 2mv\sqrt{(g/r)}$, in addition to B. Figure 4.2(b) shows the balloon shortly after liftoff. As you can see, the Coriolis force will deflect the balloon to the right and F_c will get larger and change direction as v turns to the right and gets larger and r decreases. So, if there were a ceiling (space station like a torus) it would definitely not strike directly above the launch point.

Again, if r_0 were large enough, the effect would be minimal because F_c would never get large enough to cause a significant deflection because of the $1/\sqrt{r}$ term.

Note that I have not considered another possible force, air drag. I believe this would have a small effect and would not alter the conclusions above.

Follow-up question: What would happen if an observer got into a car and accelerated beyond the rotational velocity of the cylinder in the same direction, once the vehicle lifted from the surface? Would it then glide for a certain period

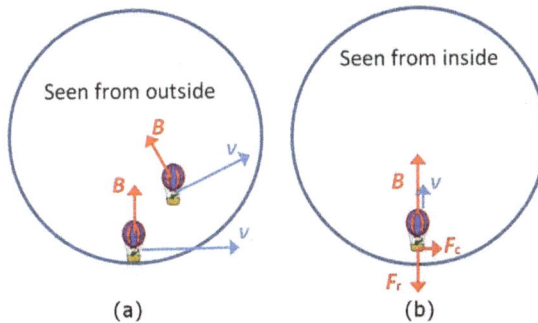

Figure 4.2. A hot air balloon rising in a rotating space station: (a) seen from an inertial frame and (b) seen in the rotating frame.

until, from the perspective of another observer inside the centrifuge, it descended back onto the surface at a slightly slower speed?

Answer: As you will see, the car will not lift off at all, so your speculations are not relevant. I will first look at the situation from inside the space station, the rotating frame. Figure 4.3(a) shows the car moving with some speed u in the same direction as the cylinder is rotating, as you stipulated. There is one real force acting on the car, the normal force N from the 'floor'. There are two fictitious forces acting, the centrifugal F_r and the Coriolis F_c, both pointing radially outward (the Coriolis force is $F_c = 2m\boldsymbol{u} \times \boldsymbol{\omega}$). So, you see, it has gotten 'heavier' rather than 'lighter' because of the Coriolis force and will not 'lift off'.

So, what happens if you travel opposite the rotation direction, as in figure 4.3(b)? Now the Coriolis force points radially inward, so there will be a speed where $N = 0$. Now, it gets a little tricky to understand what happens as u increases. The first thing you have to understand is that the car, in the frame of the rotating cylinder, is going around the circle; so the sum of all the forces must add to the centripetal force: $mu^2/r_0 = N - mg + 2mu\sqrt{(g/r_0)}$. So, when $N = 0$, $u^2 + gr_0 - 2uv = 0 = u^2 - uv$. So, when $u = v$, $N = 0$. This makes total sense because if you look at the car from outside, it is at rest and, since this is taking place in empty space, there are no forces whatever on the car. But now, since $N = 0$, the frictional force between the tires and the ground will be zero; any attempt to accelerate faster will only result in spinning wheels! So there will never be a 'lift off' situation.

This next question is interesting because it addresses the question of why objects like planets and moons tend to be spherical. In particular, it examines the gravitational field at the surface of a solid cylinder and shows how the resulting forces tend to push it toward a sphere. But exactly how such a shape change would occur depends on whether the radius is larger than the length, or *vice versa*. The mathematical details are in appendix D.

Question: I understand that, with a roughly spherical object, like the Earth, the gravitational force tends to act on objects towards the center of the sphere. What

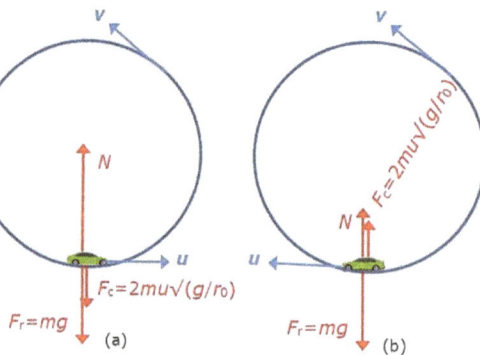

Figure 4.3. A car driving in a rotating cylinder.

would the direction of the force be with an object shaped like a cylinder? Also, would the gravitational pull be greater on each end of the cylinder than at some point in the middle? (In which case I would guess that, in nature, any massive cylinder would collapse to form a sphere?)

Answer: The gravitational field of a cylinder is pretty easy to calculate on its axis and very difficult to calculate elsewhere. Let me first provide a qualitative argument that the field at the 'poles' will be larger than at the 'equator'. At the equator, the contribution to the field from each piece of mass in the 'north' half will have a corresponding piece from the 'south' side and their axial components will cancel out, leaving only the radial components. As long as $L > R$, there will be much less cancellation for fields at the 'poles'; therefore, if the cylinder is not rigid, it will collapse to a sphere axially (from the poles; see appendix D). If $L < R$, this argument will work in the opposite way and you would expect the collapse to be radially in from the equator. At the center of each end of a cylinder of length L and radius R, the field g can be shown (see appendix D) to be $g_{end} = [GM/(RL)]$ $[(2\,L/R) + \frac{1}{2} - \frac{1}{2}\sqrt{(R^2 + L^2)}/R]$; for the case $L \gg R$, this may be approximated as $g_{end} \approx 3GM/(2R^2)$. At the equator, Gauss's law may be used to show that, for $L \gg R$, $g_{equator} \approx 2GM/(LR)$. Hence, $g_{end}/g_{equator} \approx 3\,L/(2R) \gg 1$. So, your expectation was right, but only if $R < L$. On the other hand, if $R \gg L$, the field at the pole approaches the field at the center of a uniform disk which is zero by symmetry. So, whatever the field is at the equator, the force tends to collapse the cylinder radially (inward from the equator). Figure 4.4 is a rough sketch of what the field at the surface would look like for $L > R$. I have just drawn a few field lines on the surface, but the whole field would be axially symmetric. The vectors are representative of the force you would feel if you were walking on the surface. So, walking from the equator to a pole would be like walking uphill, but you would also get heavier. If you placed a small mass on the surface, it would only be in equilibrium at the equator (stable equilibrium) or at a pole (unstable equilibrium).

The next question could just as well have been included in chapter 1, but it is a 'cool' question too.

Question: I'm trying to find a fast/easy way to test whether a sealed, consistently dimensioned rectangular box is sufficiently 'stable' for transport on a (small) two-wheeled bicycle trailer. The box is pretty tall. If it's over-weighted and top-heavy, it'll flip the trailer around turns (which are sufficiently tight/quick). I figure there might be a quick, static 'tip test' with a combination pull gauge, inclinometer and

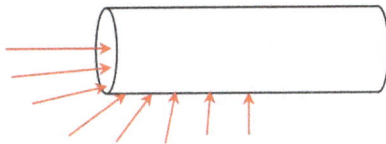

Figure 4.4. Estimated gravitational field of a uniform cylinder.

scale, but my math skills are primitive. Is there a simple way to ascertain whether, for a given object, a target stability threshold is met?

Answer: The easiest way to do this problem of your trailer turning a curve is to introduce a fictitious centrifugal force, which I will call C, pointed away from the center of the circle; the magnitude of this force will be mv^2/R, where m is the mass of the box plus trailer, v is its speed, and R is the radius of the curve. Figure 4.5 shows all the forces on the box plus trailer: W is the weight and the green x is the center of gravity (COG) of the box plus trailer; f_1 and f_2 are the frictional forces exerted by the road on the inside and outside wheels, respectively; N_1 and N_2 are the normal forces exerted by the road on the inside and outside wheels, respectively; the center of gravity of the box plus trailer is a distance H above the road and the wheel base is $2\,L$ (with the center of gravity halfway between the wheels). Newton's equations yield:

- $f_1 + f_1 = C$ for equilibrium of horizontal forces;
- $N_1 + N_2 = W$ for equilibrium of vertical forces;
- $CH + L(N_1 - N_2) = 0$ for equilibrium of torques about the red x.

If you work this out, you find the normal forces which are indicative of the weight the wheels support: $N_1 = \frac{1}{2}(W - C(H/L))$ and $N_2 = \frac{1}{2}(W + C(H/L))$. A few things to note are:

- the outer wheel supports more weight;
- if $C = 0$ (you are not turning), the inner and outer wheels each support half the weight;
- at a high enough speed C will become so large that $N_1 = 0$ and if you go any faster you will tip over;
- if the road cannot provide enough friction you will skid before you will tip over.

Now we come to your question. You first want the maximum speed without tipping. Solving for v in the $N_1 = 0$ equation gives

$$v_{max} = \sqrt{[RWL/(mH)]} = \sqrt{[RgL/H]},$$

where $g = 9.8$ m s^{-2} = 32 ft s^{-2} is the acceleration due to gravity. For example, suppose that $R = 7$ ft, $L = 17$ in $= 1.42$ ft, $H = 30$ in $= 2.5$ ft. (These dimensions were sent to me by the questioner.) Then

$$v_{max} = \sqrt{[7 \times 32 \times 1.42/(2.5)]} = 11.3 \text{ ft s}^{-1} = 7.7 \text{ mph}.$$

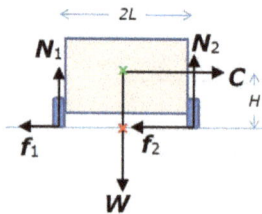

Figure 4.5. A bicycle cart rounding a curve.

Be sure to note that the assumptions of a level road (not banked) and wheels not slipping are used in my calculations. Also be sure to note that W is the weight of both the box and the trailer and $2L$ is the wheel base, not the box width.

One more thing is that you might not know how to find the COG of the trailer plus box. If the COG of the trailer is $H_{trailer}$ above the ground (probably close to the axle) and the COG of the box is H_{box} *above the ground*, then $H = (H_{box} \times W_{box} + H_{trailer} \times W_{trailer})/W$.

Added thought: When just about to tip, all the weight is on the outer wheel and so $N_2 = W$ and $f_2 = \mu N_2 = \mu W$, where μ is the coefficient of static friction. If you work it out, the minimum value μ must have to keep the trailer from skidding is $\mu_{min} = L/H$. For the example worked out above, $\mu_{min} = 0.57$. For comparison, μ for rubber on dry asphalt is about 0.9, so the trailer would not skid. (Fictitious forces were discussed in detail in book I of *Ask the Physicist, From Newton to Einstein.*)

This next question is 'cool' because by just changing one variable, the distance to the point where the rocket is released, you can visualize all possible orbits.

Question: I work in a high school where this question was posed by one of the pupils in a class I support. The question is this: if you could attach a rope to a rocket, which would also be attached to Earth, and sent it into space (out of our atmosphere) until the rope went taught and then cut the string, would it stay in space or would it fall back to Earth?

Answer: This is one of those problems which I had fun with and I hope it will not be too exhaustive an answer. I will make the following assumptions:

- the rocket always goes straight up;
- the rocket stops moving *vertically* when the rope is taught;
- the rope is cut the instant that the rocket stops;
- fuel and the weight of the rope are not issues; and
- the launch is from the equator, which makes things much simpler.

My view of this problem, therefore, is the same as if the rocket were on top of a very long stick pointing vertically straight up and the stick was suddenly removed. The thing to appreciate is that even though the rocket goes straight up, it will have the same angular velocity ω as the Earth, so its speed will be $\omega(L + R)$, where L is the length of the rope and R is the radius of the Earth. The angular velocity is $\omega = [(2\pi \text{ radians})/(24 \text{ h})] \times [(1 \text{ h})/(3600 \text{ s})] = 7.27 \times 10^{-5} \text{ s}^{-1}$. If L is just right, the rocket will assume an orbit like the geosynchronous communication satellites; this turns out to be if $L = 5.6 R$.

1. So, if the rope happens to be 5.6 times longer than the radius of the Earth, the rocket will remain (apparently) stationary above its launch point; it is actually going in a circular orbit with a period of 24 h. See figure 4.6 #1. For any other L the orbit will be an ellipse with the center of the Earth being at the focus.

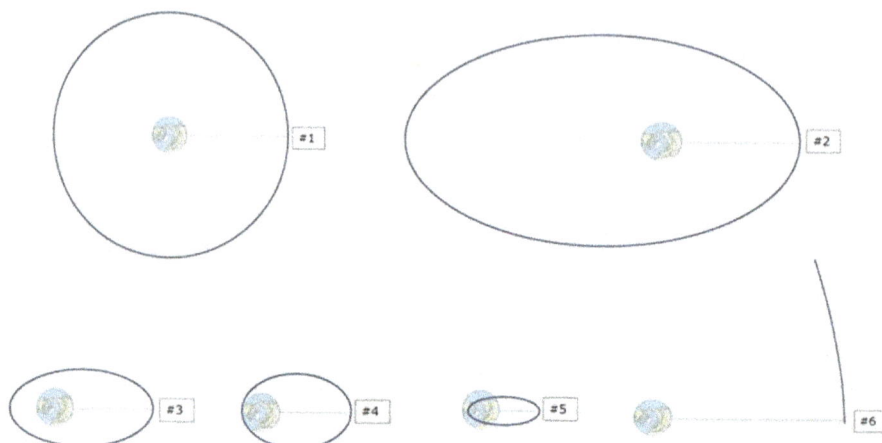

Figure 4.6. Orbits for initial distances given in 1–7 in the list.

2. If $L > 5.6\,R$, the starting point of the rocket will be the perigee (closest point to the Earth) of its orbit. See figure 4.6 #2.
3. If $L < 5.6\,R$, the starting point of the rocket will be the apogee (farthest point from the Earth) of its orbit. See figure 4.6 #3.
4. For $L < 5.6\,R$, though, there will be some critical distance $L = L_c$ when the perigee of that orbit is exactly equal to R; in that case the orbit will just skim the surface of the Earth. After some laborious algebra I found that $L_c \approx 3.7\,R$, about 1.9 Earth radii inside the geosynchronous orbit. See figure 4.6 #4.
5. For $L < L_c$, the rocket will crash back into the Earth, but not where it was launched from because it is a projectile which has a horizontal speed that is greater than that of the Earth's surface. See figure 4.6 #5.
6. Finally, if the horizontal speed of the rocket $\omega(L + R)$ is greater than or equal to the escape velocity, $v_e = \sqrt{[2MG/(R + L)]}$, where M is the mass of the Earth, the rocket will escape the Earth and never come back. I calculated this to be when $L = 7.6\,R$, two Earth radii beyond the geosynchronous orbit. See figure 4.6 #6.

The next question is from a photographer who is experimenting with pinhole cameras. He has been studying the physics of these cameras, specifically how to choose the hole diameter for optimal resolution of the image. The second half of this question gets a little more mathematically technical than my normal answer.

Question: I am a photographer and general tinkerer. I have started building pinhole cameras and during the course of building them I came across some equations on obtaining the optimal pinhole size. I find I understand the equations well, but I cannot find a solid explanation for the 2.44 constant used in the calculation for the Airy disk (2.44 × light wave length × focal length). Everything I have read is very vague and I want an explanation. Why do we use this number?

I read that it is like pi, it just is, but is to what, waves? Is it a speed? What does it represent? I want to understand.

Answer: This verges on being too technical for this website, but I always want to encourage folks to 'have an enquiring mind', so I will take a stab at it. This question requires that you understand a little about diffraction, the interference of light waves. If we shine light through a slit and the slit is very large compared to the wavelength of the light we are looking at, we get an 'image' of the slit on a screen. If the rays of light hitting the screen come from very far away compared to the size of the slit, the image of the slit will be the same size as the slit itself. The Sun coming in through a window creates an image of the window, the same size, on the wall. So, I now want to make images of smaller and smaller slits; what happens is that we come to a point where as we make the slit smaller, the image starts getting bigger. Figure 4.7(a) shows what the 'image' of a very narrow slit looks like (bottom) and a graph of the intensity (top). It is fairly straightforward to show (see any elementary physics textbook) that $a\sin\theta = \lambda$ gives the angle of the first dark spot in the pattern for wavelength λ; the geometry showing a and θ is shown in figure 4.7(b). Now, a pinhole is not a slit, but it is very similar and you would expect its pattern to be very similar. Indeed, instead of getting stripes you get a bullseye pattern. Doing the analysis in a similar way to that which leads to the angle of the first minimum gives a slightly different result, $a\sin\theta = 1.22\lambda$, where a is the diameter of the hole. In essence, this is where your factor of 2.44 comes from. But, specifically, where does it come from? It gets pretty technical here! When you solve the pinhole diffraction problem, you work in cylindrical coordinates because the problem has cylindrical symmetry. As is often the case with problems with this symmetry, the solution (for the intensity in this case) involves a Bessel function, a special mathematical function which whole books have been written about. The intensity is given by $I = I_0[J_1(\frac{1}{2}ka\sin\theta)/(\frac{1}{2}ka\sin\theta)]^2$, where J_1 is the first order Bessel function and $k = 2\pi/\lambda$. Now, we are interested in when the intensity is first zero; the first zero of $J_1(x)$ is for $x = 3.832$, so $\frac{1}{2}ka\sin\theta = (\pi a/\lambda)\sin\theta = 3.832$, or $a\sin\theta = 1.22\lambda$. Next, we convert the angle to lengths by

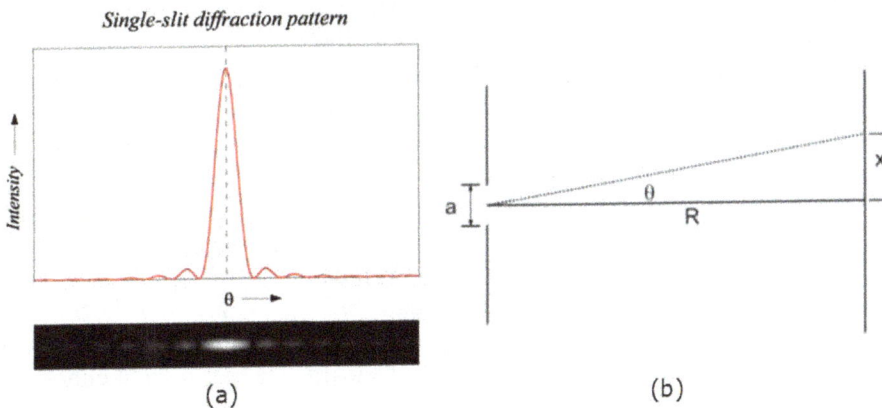

Figure 4.7. (a) Single-slit diffraction pattern and (b) the geometry.

approximating (see figure 4.7(b)) $\sin \theta \approx x/R$ so $x = 1.22R\lambda/a$ or $2x = 2.44R\lambda/a$; $2x$ is the diameter of the smallest spot to which a collimated beam of light can be focused. Since you say that you 'understand the equations well', except where 2.44 comes from, I guess my task is complete. Apparently (from the little research I did), the optimal size is if the minimum spot size equals the hole size, i.e. $2x = a_{opt}$, so $a_{opt} = \sqrt{(2.44R\lambda)}$. There is an alternative form of optimum size which is based on the Rayleigh criterion for resolving two spots (which stipulates that the central maximum of one image is on the first minimum of the second), which has $a_{opt} = \sqrt{(3.66R\lambda)}$.

Sometimes my answer to a question depends not on my awesome ability to do physics calculations, but rather on using the internet to do a little research. The following Q&A is an example where I came in with some notion of how things are and was startled to find that my notion was not even approximately correct. Nearly everyone who has ever taken an introductory physics course, and certainly everyone who has ever taught one, is familiar with the problem where you imagine drilling a hole all the way through the Earth and dropping something in it. If you model the Earth as a uniform sphere the force turns out to be the same as for a mass on a spring—the force is toward the center and proportional to the distance from the center. So, neglecting any frictional forces, something dropped into the hole will oscillate back and forth between the two hole openings; it turns out that the period is the same as a near-Earth orbit, about an hour and a half. It also turns out (unsurprisingly, if you think about it) that calling the Earth a uniform sphere is really a poor model. I love doing *Ask the Physicist* because I learn something new almost every day!

Question: Have scientists done an experiment on what the value of gravity is below the Earth's surface as depth increases? If so, please provide a chart of g vs depth.

Answer: The deepest hole ever drilled is only about 12 km deep. I could not find any reference to attempts to measure g at various depths down this hole. Since the radius is about 6.4×10^3 km, you would only expect about a 0.2% variation over that distance. There are models of the density of the Earth, though, which have been determined by observing waves transmitted through the Earth during earthquakes or nuclear bomb tests; these are believed to be a pretty good representation of the radial density and can be used to calculate g. Figure 4.8 shows the deduced density distribution and the calculated g. Usually in introductory physics classes we talk about the Earth having constant density, but, as you can see, that is far from true—the core is much more dense than the mantles and crust. If it were true, g would decrease linearly to zero inside the Earth. Instead, it first increases slightly to around 10 m s^{-2} and then remains nearly constant until you are at a depth of around 2000 km. There is little likelihood that g will ever actually be measured deep inside the Earth because the temperature

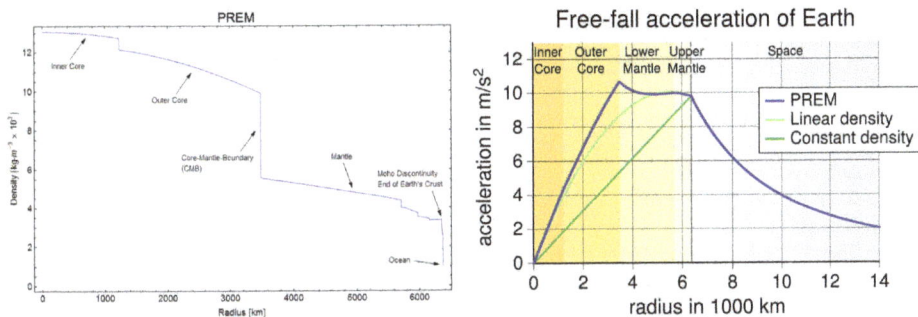

Figure 4.8. Left: density inferred from seismic data (PREM). Right: g for PREM and simpler densities.

increases greatly as you go deeper, and is already nearly 200 °C at 12 km. However, if you have detailed information on density distribution, there is really no need to measure g.

The next question is, I think, particularly cool. The more I think about human-kind venturing to places more distant than our solar system, the more obstacles I can see to extremely high-velocity travel. One that occurred to me is illustrated in this question.

Question: A spacecraft traveling at very near light speed sets a course for a distant object. So, traveling at near light speed, could a navigation computer make corrections quickly enough, and alter the trajectory for objects in its path. Could these calculations be made quickly enough to miss the object? Assuming that computer calculations happen at the speed of light, the craft could be on a collision course with an object (a) before it saw it and (b) before it could recalculate an evasive action to miss the object. There would be the added disadvantage that human thought isn't anywhere near quick enough to react to a blocking object. So how could 'we' travel to distant objects without fear of plowing into the first path-crossing object?

Answer: You raise an interesting and important question here. But it is much worse than you think. For one thing, because of length contraction, an object which an outside observer would see as being, for example, one light year away from you would only be $1 \times \sqrt{(1 - .99^2)} = 0.14$ light years away as you observed it when going 99% the speed of light; that gives you a lot less time to maneuver. But there is a much more important barrier to being able to navigate at very high speeds. The upper picture in figure 4.9 (left) shows the spacecraft as seen by an outside observer; light from a distant star comes with velocity c, making an angle of θ relative to your velocity v. But, if you now are on the spacecraft, you observe the star at a different location specified by θ' because of the velocity trans-formation to the new light ray c'. Note that the magnitudes are the same, $c' = c$, but the directions are different, $\theta' \neq \theta$. It can be shown that $\theta' = \tan^{-1}[\sin\theta\sqrt{(1 - (v/c)^2)}/(\cos\theta + (v/c))]$. I have plotted this for several values of v in figure 4.9 (right).

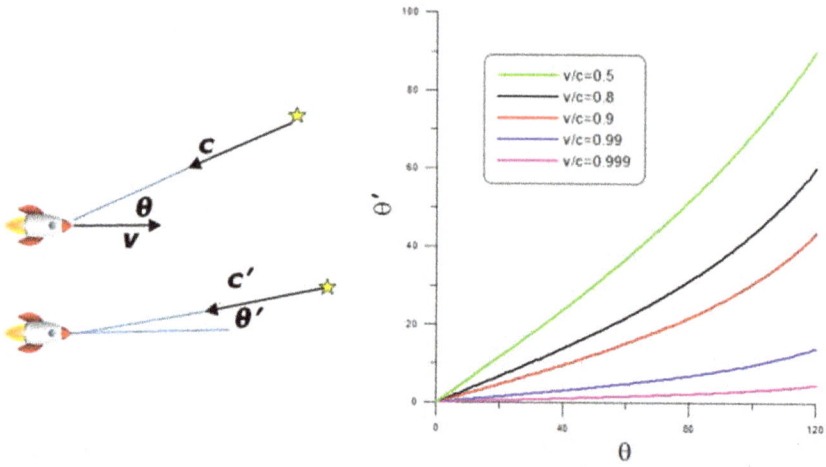

Figure 4.9. Left—above is the outside frame and below is the spacecraft frame. Right—the observed angle as a function of the rest-frame angle for several speeds.

The effect is quite dramatic. For example, at 99.9% the speed of light everything in your forward hemisphere and much from behind you will appear inside a 5° cone in front of you! Just having optics which could give good enough resolution to see anything but an extremely bright light in front of you would be a challenge in itself. But wait, it gets even worse! There will be significant Doppler shifts and much of the light at very high speeds will be shifted out of the visible spectrum.

Chapter 5

Physics is frivolous

5.1 Introduction

I often get questions which seek to find the physics of 'frivolous' situations, situations involving superheroes, video games, science fiction, etc. These are fun, and they often lead to outrageously impossible physics from the viewpoint of practicality.

5.2 Frivolous physics

Question: Two of us disagree on part of a solution given by two people with physics backgrounds, and I want to know if I am correct, or if I am missing something in the analysis of the problem, in case I have to explain it to a student. The question concerns forces/impulse. A 50 kg person falling with a speed of 15 m s^{-1} is caught by a superhero, and the final velocity up is 10 m s^{-1}. Find the change in velocity. Find the change in momentum. It takes 0.1 s to catch them. What was the average force? The answers are: Person A says the change in velocity = 25 m s^{-1}, the change in momentum = 1250 kg · m s^{-1}, and the average force is 12 500 N. Here's where we disagree. Person B says that 12 500 N is equivalent to 25*g*. They try to explain that 250 m s^{-2} acceleration corresponds to 25*g*. I said this makes no sense at all—I know the acceleration is 250 m s^{-2}, but that doesn't in any way imply a 25*g* 'equivalence' to me. Person B then went further to 'prove' their point. Here is their argument. 500 N/g = 12 500 N/n*g*. I agree with the *n* = 25, but say there is no justification for the 500 N/g in the first place. Do you have any ideas about where it comes from, or how to justify that value? By the way, I teach physics on and off at the high school level. Person B is an engineer, I think.

Answer: Person B is wrong but has the right idea. (As you and your friend have apparently done, I will approximate $g \approx 10$ m s^{-2}.) We can agree that the acceleration is $a = 250$ m s^{-2} and that is undoubtedly 25*g*. Now, we need to write Newton's second law for the person, $-mg + F = ma = -500 + F = 12\,500$, so $F = 13\,000$ N. This is the average force by the superhero on the person as she is stopped, so the answer that the average force is 12 500 N is wrong. When one expresses a force as 'gs of force', this is a comparison of the force F to the weight of the object mg, F(in gs) = F(in N)/mg = 13 000/500 = 26g; this simply means that the force on the object is 26 times the object's weight. So neither of you is completely right, but if there is any money riding on this, your friend should be the winner because the only error he made was to forget about the contribution of the weight to the calculation of the force. I am hoping that superman knows enough physics to make the time be at least 0.3 s, so that Lois does not get badly hurt!

The following question is a perfect example of how totally impossible something can be, from a 'where is the required energy going to come from?' point of view.

Question: In some of the more realistic space combat in science fiction there is a concept called a 'BFR' (big fast rock), in which matter is mined from a dead world or asteroid, melted to molten and then reformed to a near-perfect density distribution with collars of ferrous metal impressed in it, before being shot at some fraction of light speed from a large EML cannon running down the long axis of mile-long ships. I would like to know how to calculate the impact force release for a 2 000 lb BFR moving at 0.10, 0.15, and 0.30 c. I'm assuming it's going to be in the high megaton range and I don't know what the translative per ton equivalence is in TNT.

Answer: What you want is the energy your projectile has when it hits. ('Impact force release' has no meaning in physics.) The energy in joules is $E = m\gamma c^2$, where $\gamma = \sqrt{(1 - (v/c)^2)}$. In your case, 2000 lb = 907 kg, $c = 3 \times 10^8$ m s^{-1}, $\gamma = 1.005$, 1.011, and 1.048. The energies in joules are $E = 8.204 \times 10^{19}$, 8.253×10^{19}, and 8.555×10^{19} J. There are about 4×10^9 J per ton of TNT, so the energies are 20.51, 20.63, and 21.39 megatons of TNT. I might add that this is not actually very 'realistic'. Where are you going to get that much energy (you have to supply it somehow) in the middle of empty space? Or, look at it this way: I figure that for a mile-long gun the time to accelerate the BFR to 0.1c is about $t = 10^{-4}$ s. During that time the required power is about $8 \times 10^{19}/10^{-4} = 8 \times 10^{23}$ W = 8×10^{14} GW; the largest power plant on the Earth is about 6 GW! Also, don't forget about the recoil of the ship, which would likely destroy it. I am afraid that I would have to label your BFR as completely unrealistic! Are you sure that BFR stands for big fast rock?

The next question addresses an important question that is often ignored in science fiction—the effect of acceleration on humans. The maximum acceleration which can be endured, and only for short times, is approximately 8g. In other words, 'jump to light speed, Scotty' would crush everybody on the *Enterprise*. So, getting up to speeds near c would take a pretty long time because we can only really be comfortable near g. How long is long? The positive side of this is that the acceleration of the ship will provide artificial gravity.

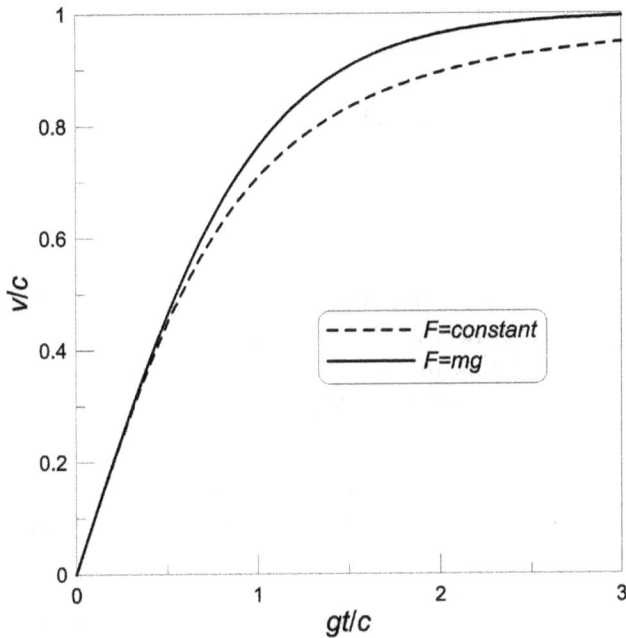

Figure 5.1. Speed as a function of time for a constant force and for a force *mg* using the velocity-dependent mass.

Question: A starship pilot wants to set her spaceship to light speed, but the crew and passengers can only endure a force up to 1.2 times their weight. Assuming the pilot can maintain a constant rate of acceleration, what is the minimum time she can safely achieve light speed?

Answer: This question completely ignores special relativity. It is impossible to go as fast as the speed of light. Furthermore, acceleration is not really a useful quantity in special relativity and you must use special relativity when speeds become comparable to the speed of light. I have previously worked out the velocity of something which would correspond to the occupants of your spaceship experiencing a force equal to their own weight due to the acceleration, which I will adapt to your case later (see figure 5.1.) First, though, I will work out the (incorrect) Newtonian calculation that is presumably what you want. The appropriate equation would be $v = at$, where $v = 3 \times 10^8$ m s^{-1}, $a = 1.2g = 11.8$ m s^{-2}, and t is the time to reach v; the solution is 2.5×10^7 s $= 0.79$ years. For the correct calculation, you cannot reach the speed of light; from figure 5.1 (full-drawn curve), though, you can see that you would reach more than 99% of c when $gt/c = 3$. To make this your problem, we simply replace g by $1.2g$ and solve for t. I find that $t = 7.7 \times 10^7$ s $= 2.4$ years, about three times longer than the classical calculation. (Note: the calculations for figure 5.1 are explained in *From Newton to Einstein*, the first volume in the *Ask The Physicist* series.)

The next question is short and sweet and the answer is even shorter! When we reach 'transgalactic civilization status', we will have to seek some other energy source!

Question: Would it be possible physically to encapsulate a black hole with solar panel type devices and use its energy to power a civilization? Like when we reach transgalactic civilization status and run across a black hole and want to utilize it. **Answer:** Well, that is a pretty crazy idea because a black hole is an energy sink, not an energy source!

The next question is about how fast the superhero Flash can run in the real world. Flash is a character(s) in DC comics who has somehow acquired the superpower to run really fast; I had never heard of him, but there are limitations in the real world, even for superheroes! (I think this was probably a homework question, a no-no at AskThePhysicist.com! I guess this one slipped past me.)

Question: Using real-world estimates for the coefficient of friction between his feet and the ground, how fast could the Flash run a quarter-mile? Assume that the limiting factor for his acceleration is the force parallel to the ground that his feet can apply.
Answer: Suppose he is running on a dry asphalt road with rubber-sole shoes. Then the coefficient of static friction is approximately $\mu \approx 0.8$. The maximum force of friction on level ground would be $f_{max} \approx \mu N = \mu mg \approx 8\,m$, where m is his mass. So, his acceleration would be $a = f_{max}/m = 8$ m s^{-2}. A quarter mile is about 400 m, so assuming uniform acceleration, the appropriate kinematic equation would be $400 = \frac{1}{2}at^2 = 4t^2$, so $t = 10$ s. His final speed would be $v = at = 80$ m s^{-1} = 179 mph.

The next question is not really so 'frivolous', but also involves static friction. We tend to think that the limit to how hard we can push something depends on how strong we are. But the real limit is friction. For example, pushing a stuck car on an icy day often ends with you down on the ground, having slipped on the icy road while pushing.

Question: How strong would a man have to be to push a 16 000 lb bus on a flat surface?
Answer: That depends on how much friction there is. And not just the friction on the bus, but, more importantly, the friction between the man's feet and the ground. Newton's third law says that the force the man exerts on the bus is equal and opposite the force which the bus exerts on the man (**B** in figure 5.2). Other forces on the man are his weight (**W**), the friction the road exerts on his feet (**f**), and the force that the road exerts up on him (**N**). If the bus is not moving, $N = W$ and $f = B$, equilibrium. The biggest that the frictional force can be without the man's feet slipping is $f = \mu N$, where μ is the coefficient of the static friction between the shoe soles and the road surface. A typical value of μ for rubber on

Figure 5.2. Strong man pushing a bus. Copyright: Jennifer Gottschalk/Anabela88 Shutterstock.

asphalt, for example, is $\mu \approx 1$, so the biggest f could be is approximately his weight W; this means that the largest force he could exert on the bus without slipping would be about equal to his weight. Taking $W \approx 200$ lb, if the frictional force on the bus is taken to be zero, the bus would accelerate forward with an acceleration of $a = Bg/16\,000 = 200 \times 32/16\,000 = 0.4$ ft s^{-2}, where $g = 32$ ft s^{-2} is the acceleration due to gravity; this means that after 10 s the bus would be moving forward with a speed 4 ft s^{-1}. If there were a 100 lb frictional force acting on the bus, the acceleration would only be $a = 0.2$ ft s^{-2}. If there were a frictional force greater than 200 lb acting on the bus, the man could not move it.

I have no idea where this next question came from, but it has the added interesting aspect that we need to calculate the total energy to gravitationally disassociate a uniform sphere.

Question: What would the yield of a 5000 ton iron slug accelerated at 95% of c by, say, a bored omnipotent be? Would it be enough to mass scatter a planet?
Answer: I get the strangest questions sometimes! So, 5000 metric tons $= 5 \times 10^6$ kg. The kinetic energy would be $K = E - mc^2 = mc^2[(1/\sqrt{(1 - .95^2)}) - 1] \approx 10^{24}$ J. The energy U required to totally disassemble a uniform mass M of radius R is $U = 3GM^2/(5\,R)$, where $G = 6.67 \times 10^{-11}$ is the universal gravitational constant. So, taking the Earth as a 'typical' planet, $U = 3 \times 6.67 \times 10^{-11}(6 \times 10^{24})^2/(5 \cdot 6.4 \times 10^6)$ $\approx 2 \times 10^{32}$ J. So your god's slug is far short of supplying enough energy to totally blast apart the Earth.

The next question again illustrates how, so often, the awesome weapons which play such an important role in video games have as their biggest problem the impossibility of the power source.

Question: I was playing a game known as *Fallout* 3 and in the game there are laser weapons. The laser weapons are powered by Marshmallow sized micro-fusion cells that are basically miniature nuclear reactors that fuse hydrogen

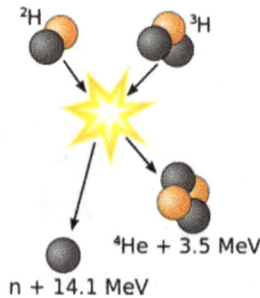

^{2}H

^{3}H

$^{4}He + 3.5$ MeV

$n + 14.1$ MeV

Figure 5.3. A fusion reaction.

atoms. In the game they produce enough power to turn a 500 kilogram bear into ash in one second. So, could a reactor that small produce enough power for the gun and how much energy would a Marshmallow-sized blob of fused hydrogen produce? A normal microfusion cell in the game has enough energy to fire 24 of these shots. Would it be possible in any way for these laser weapons to be able to be this powerful with an energy source like the microfusion cell?

Answer: I have no way to estimate the 'power to turn a 500 kilogram bear into ash in one second'. I am sure you realize that, with today's technology, the possibility of there being such a power supply is zilch. Let's just do a few estimates to show how hard this is. One gram of hydrogen fuel (deuterium + tritium), if fully fused into helium + neutrons, figure 5.3, releases something on the order of 300 GJ of energy; so, if released in 24 one second pulses, each pulse would be about 10 GW. That is probably way more than your bear burning would need, so let's say 100 MW would do it; so, we would need about 10^{-2} g of fuel. I calculate that to confine that amount of gas in a volume of 10^{-5} m^3 (about 1 in^3) would require a pressure of about 5 000 000 atmospheres! That, in itself, should be enough to convince you that this machine could probably never be possible. If you need more convincing, consider shielding: 80% of the energy produced is in the kinetic energy of neutrons. How are you going to harvest that energy in such a small volume and how are you going to protect yourself from the huge neutron flux? And surely there needs to be some sort of mechanism to control the process and convert the energy into usable electrical energy to power the laser; all that is supposed to fit into 1 in^3? This truly is a fantasy game with no connection to reality!

The next question, from a sci-fi screenwriter, proposes using linear acceleration to create artificial gravity during a trip from the Earth to the Moon. This will get you there very quickly!

Question: I am a writer putting together a science fiction screenplay. Those who know me say I have an attention to detail—to a fault. There is one particular element I would like to be as accurate as possible. I'm hoping you might be able to help me. Here is the scenario. A spacecraft leaves Earth on course to the Moon. In order to create an Earth-like gravity inside the ship, the ship accelerates at a constant rate, exerting a force on the occupants equal to one g. Halfway through the trip the craft will flip, then decelerate for the remainder of the journey. This would give the same sensation of false gravity to the occupants of the craft. So here is the question: if this were possible; how long would it take to actually reach the Moon?

Answer: Since you are such a stickler for detail, I will give you detail, probably far more than you want! Your scheme of having an acceleration with 'constant rate' would work in empty space, but not between the Earth and the Moon because the force causing the acceleration is not the only force on you, the Earth's and the Moon's gravity are also acting. As you go away from the Earth, the Earth's gravity gets smaller like $1/r^2$, where r is the distance from the Earth's center, and the Moon's gets bigger as you get closer. So, it becomes a complicated problem as to how much force must be applied to keep the acceleration just right for where you are. Let's call your mass M. Then there are two forces on you, your weight W down and the force the scale you are standing on exerts on you, F. W gets smaller as you get farther and farther away and you want F to always be what your weight would be on the Earth's surface, Mg. So, Newton's second law says that $F - W = Ma = Mg - W$, where a is the acceleration you must have. Note that, for the time being, I am ignoring the Moon; that would just complicate things and its force is much smaller than the Earth's, at least for the first half of the trip. I want you to understand the complication caused by the fact that W changes as you go farther away. Now, how does W change? $W = MM_E G/r^2 = Mg (R/r)^2$, where R is the radius of the Earth and M_E is the mass of the Earth. We can now solve for the acceleration the spacecraft would have to have: $a = g(1 - (R/r)^2)$. I have plotted this in red in figure 5.4. (The distance to the Moon is about 60 Earth radii.) Note that for most of the trip the acceleration is just about g. I also calculated the effect the Moon would have (blue dashed line) and, except for the very end of the trip, it is pretty negligible. Now that we have taken care of the always-important details, we can try to answer your question. To calculate the time exactly would be very complicated, but, since the required acceleration is g for almost the whole trip, it looks like we can get a really good approximation

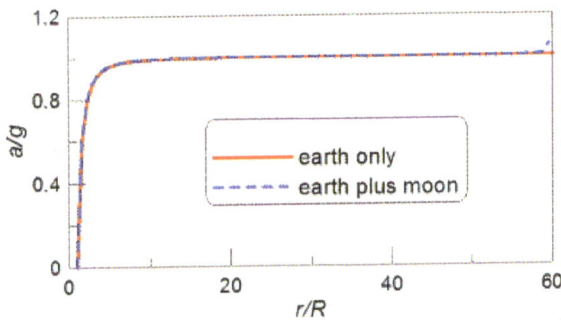

Figure 5.4. Required acceleration for the net force to be equal to W on Earth's surface.

by just assuming $a = g$ for the first half and $a = -g$ for the second half; your perceived weight (F) will just decrease from twice its usual value when you take off to about normal when you get to about five Earth radii in altitude. The symmetry of the situation is such that I need only calculate the time for the first half of the trip and double it. The appropriate equation to use is $r = r_0 + v_0 t + \frac{1}{2}at^2$, where $r_0 = R$ is where you start and $v_0 = 0$ is the speed you start with. Halfway to the Moon is about $r = 30\,R = R + \frac{1}{2}at^2$ and so, putting in the numbers, I find $t \approx 1.69$ hours, and so the time to the Moon would be about 3.4 hours. You can also calculate the maximum speed, you would have to be about 140 000 mph halfway.

Note that the calculation I did above to generate figure 5.4 was the acceleration if you continued speeding up the whole way. I should have had the acceleration switch to (approximately) -9.8 m s^{-2} halfway so as to slow down. But the important takeaway here is that once you are more than a few Earth radii on your way, you can ignore the Earth altogether as the questioner did for the whole trip.

Chapter 6

Physics isn't...

6.1 Introduction

There are three admonitions to greet visitors to AskThePhysicist.com. One is that homework problems or tutoring assistance are strictly forbidden. No matter how vociferously I proclaim that these are not the purpose of the site, probably 95% of all questions fall into this category, and I find them very annoying. The other two, however, are less annoying because they provide some entertainment and levity— questions from folks who think I am a psychic and questions from armchair physicists who have some crazy theory about how the Universe works.

6.2 Physics isn't psychics

It is amazing how many visitors to the *Ask the Physicist* website think they are asking the psychic. In spite of the disclaimers I have put on the homepage, nearly every day I receive at least one question asking me to use my mystical powers to help guide their daily lives or explain remarkable phenomena. And, it is not just ignorant people who cannot spell—computer algorithms don't seem to get it either. A few years ago I tried to use GoogleAds on the site to generate a little money; to my horror, many ads generated by Google's software were for 'Free Psychic Readings'! And, if you have ever tried to talk to a human being at Google, forget it; I just gave up using GoogleAds when I couldn't stop the psychic ads—what better way to destroy the credibility of a serious science site than with ads for palmists and clairvoyants? Gathered here are some of my favorites from Ask the Psychic. (By the way, unlike the serious parts of this book, questions in this and the next section are not edited.)

Question: I'm going to place a little task here and see how you can crack it.. If you can solve it, get to me by using the medium
Question: Found a small bottle of sand and fine stone tied to the floor of my second hand car. What does it mean

Question: I wanted to know if my ex Jorge Rodriguez still loves me? Will he talk to me soon?

Question: Hello, i have lost my phone and i really need it as soon as i can any hints or clues on where to find it? Thank you, Love Josh

Question: Besides hearing marbles on a rope all over my roof I came back frm store gone for an hour went to pee as soon as I came inside n my electric heater was n a puddle of water wow!

Question: Will I pass my drug test

Question: I split up with my husband 7weeks ago since then I found out he was having an affair and is now living with her. He's called Mark Wright dob 1.2.1972 is he happy with her or will they split up and he return to me

Question: Will i ever be famous

Question: am i gonna get killed? i DONT WANNA DIE plz awnser i love you!

Question: Hello Two of my friends seems to be avoiding me. Maybe they don't want to be friends anymore? I don't know, the whole thing just weird. I tried to speak to them about me feeling like an outsider, but one of them just got angry and blames it all on me. The other one didn't even answer. I'm the one always taking contact, and i'm just really scared of asking them «don't they want to be friends anymore?»

Question: Will Rori be okay? can I save her? is mummum still here?

Question: Will I have another baby by Poppa?

Question: do i love akhter balouch? me jabeen

Question: Will I have a wife and kids before I die

Question: i was wondering if this girl madeline scott has a crush on me.

Question: how can i get a boyfriend

6.3 Physics isn't crazy

Although I emphasize that I require single, concise, well-focused questions, inevitably I get lengthy questions (dissertations, really) from people wishing to get the stamp of approval of a professional physicist on their personal theory of something or other. If these submissions are not questions but rather assertions, then they are start with something like 'Is it possible that...?' Of course a scientist is often loathe to say that something is impossible, even if he believes it to be—then you get attacked for being closed-minded. Best to just file these away in my Off-the-Wall Hall of Fame page rather than try to get into a dialog with a crazy person! Here are a few good ones.

Question: would it be possible to make a laser gun by taking Plutonium or uranium and putting the material in a container with a pipe in the middle and the radio active material around the outside of the inner pipe, then have a laser pass through the radiation field it give off because it gives off heat. would it increase the lasers power so it would be strong enough to burn through metal or shoot someone like the fazers in star track. I want to be the first to figure out how to make a gun that would be able to shoot in space, because the bullet of gun would need a lot of power behind it to propel forward in the vacuum of space.

Question: I believe I have theorized a way to create an Alcuberie drive without using generally unobtainable (as of today) materials. I am no expert physicist, I am only 17 and have not taken any college yet. I just watched a documentary on gravitational waves, just out of curiosity, and I had been working on designing a plausible warp drive system. the main problem was the negative energy induced spacetime curve. now imagine having a constant spacetime well in the front of the ship. we could (possibly) cause this by spinning a dense wheel to incredible speeds ($100\,000$ m s^{-1} +), whilst holding it together using electromagnets, and dissipating any heat generated using a high surface area metal shape which emits light to cool down. According the Lorentz factor equation, this should work. But now imagine the same wheel, on the back of the ship, with lots of weights on spokes from the center, with proportional distance between them to produce a sine-ish waveform relative to every point in the weight's path. this waveform occurs just like how a water drop into a pool causes a peak and a trough of a wave. Here is the main idea. We have the wheels spinning at the same rate. when the wave in any given point in the circle is at a trough, there will be an equal spacetime disturbance in the front of the ship. No warping occurs. But magic happens when the peaks of the wave occur, as a reverse effect of the trough. Since the spacetime at the peak of the wave is accelerated in time, and the front is slower in time, the requirements for the alcuberie drive's spacetime curve are met. this theoretically produces a warp bubble around the ship, causing it to go faster than the speed of light if the wheel speed is fast enough.

If you think this would work or I should contact nasa or someone else, just let me know. I also have tons of other new technological ideas such as a kinetic based gyroscopic propulsion system, hydrazine powered cars, and some other things, as well as theories. Hope to hear from you soon.

Question: So ... if im at the center of a gyroscope ... free of my own orientation ... and say this gyroscope is creating an electrical spere around me ... and the center of this electrical spere the charge is zero ... thus not effecting me ... motuon has no effect ... no g forces are felt upon motion ... no natter the speed ... would i not then be able to coordinate a lat and long destination ... griund my electrical charge to that destination ... and shoot my gyro like abolt ... having a gold ball on a spike on one side and a non conducive ball at the other ... traveling at the speed of light by my gyro generating such a great electrical charge from the motion of its rings ... make since?

Question: Could the mysterious DARK ENERGY be the accumulated photons emitted from stars and other bodies over the last 14 billion years?

Fact: Astrophysicist's are always fascinating us with pictures of galaxies that are 'billions of light years' away. What that means in a practical sense, is that the photons of light captured for that snapshot were freely roaming the Universe for billions of years, and were part of the Universe's total energy. Now between the emission of that photon from a billion years ago till the emission of today's photon from that same galaxy (assuming it is still there), the Universe accumulated another billion years worth of photon emissions from each and every star in that galaxy. Each photon accumulates in the free space of the Universe and mixes with each other photon in the Universe that may be passing the same space. So, I'm thinking that all these free photons mixed together form a soup or foam of energy, completely unseen, but still existing.

Proof of their existence is the photograph of the Galaxy 'billions of light years distance' from us. The photons we collected on the photo plate were emitted a billion years ago and still exist.

Also, if what I posit is correct, and photons are the DARK ENERGY, than couldn't we explain the acceleration of the expansion of the Universe is caused by the ever increasing amount of photons accumulated in the Universe by the constant emission of more photons by the stars.

It would be like the total photon energy of the Universe is constantly increasing because of new photons added to the mix.

More energy equals acceleration?

Question: Is the Earth spinning because the Sun is a negative charge resting on the circumference of the Earth and that space is a positive energy source also pushing on the other pole? much like a metal ball being suspended and stabilized with positive negative charges? if so then would or could the Earth be spinning much like a top does in a circular motion as if it will always spin in the same spin motion of earth because it to is affected by the magnetic pushes and pulls ... right? could we measure whether the Earth is spinning at the same rate of speed as it was 10, 15, 25, or even 50 years ago? if it is slowing does that mean that if earth loses momentum this will cause the Spheres to be slowly deteriorated mime-ing global warming? could we feasibly measure the rate at which earth would stop rotating if at all? does it mimmick the pi equation at all? I am sorry I have only a couple of years of college and I have been bothered buy these questions ... . I am fifty years old and an introvert I look forward to your answer

Question: Working on model for my Theory of The Dark Process, Jason D. Lispi, and now deep in the math. To consider e or q, in either of their respective and functional arenas, a 'true' Constant based on give and takes, such in the case of Elementary Charge, based on my Charge cojagation revalues where something like the transitional phases are NOT a specific point of observation. In my math, the transitions, even partially assessed Physically, analogously but with a $-n(n)$ being expressed seems to me, obviously, 100% accurately compliant within my theory as a TRUE CONSTANT. How did we come to this q thing to be a 1: 1/3 (3)to = 1.Just another Shot Noise tool. THIS IS AWEFUL. I have such a higher degree of truthful continuous continuity inclusive is ... . Well I have started at before the Big Puncture CME (Dark Nature) and followed the leak through it's transitional phases and have only Borderless Boundary slowing me at the 'massive Bodies' in motion level. There seems a distribution method is applied to these areas. It seems to be a distributive sharing of sorts, all repulsive, yet to equate to a outbound repulsion force very close in nature to quantities, perfect enough to abate the inbound and always intruding, Dark Force, repulsive and part of the Dark Process. This system is very constant yet ACTIVELY and logically endorses and applies adjustments in this 'sharing' or 'distributing' sub process for orbital and spiral mechanics to flow well and predictably. Avoiding a big crunch and possible? CME. I would love to share my work, who do you recommend?? Please advise. I am only a Janitor. TYVM. 'The Dark is the Light.' Time to see it!

Question: I am not a student and have little to no knowledge of this subject. Still after reading some on dark matter I was compelled to to fitnd someone who'd know the answer to my question about that 20%. So my question is this. In caculating earth's mass. When determining the 20% discrepancy that we believe to be dark matter. Was human and an animal mass also included? However if my last question was with out any sense since the missing 20% (dark matter) Is within the solar system and not in the mass of earth alone. Still earth's mass I would believe must contain the living mass as well. This missing 20% may pertain to the mass of unknown living mass throughout out the solar system. Supporting to what of angels and demons also called aliens, spirits, god's, etc. I do hope that my questions are not completely idiotic. I just felt that for some reason this was something that was not considered when determining mass of earth since living mass is always changing due to population. How ever when u consider all life come from the father and he always was and always will be. Then his mass would be likewise. In that in which he breathed into the living. True?

Appendix A

The constants of electricity and magnetism

Most textbooks introduce the unit of charge (coulomb) before the unit of current (ampere). I have done it a little differently here because the thing that is actually operationally defined is the ampere. Since this book does almost no electricity and magnetism, there is no real reason to introduce the coulomb first. In addition, most laypersons have a much better idea of what an ampere is than they do about what a coulomb is.

Two long parallel wires, each of length L and separated by a distance r carry electric currents I_1 and I_2. They are observed to exert equal and opposite forces (Newton's third law) on each other and the magnitude of this force is proportional to the each of the currents and the length of the wires, and inversely proportional to the separation: $F \propto LI_1I_2/r$. Choosing a proportionality constant $\mu_0/(4\pi) = 10^{-7}$ defines what the unit of electric current is: $F/L = \mu_0 I_1 I_2/(2\pi r)$. So, if two wires carrying equal currents are separated by 1 m, and the force per meter each wire experiences is 2×10^{-7} N m^{-1}, then each wire is carrying an electric current of 1 ampere (A). Since electric current is the rate at which electric charge is flowing, knowing the ampere also lets us know the electric charge unit we will use, called the coulomb (C), because 1 A = 1 C s^{-1}. To set the scale relative to everyday life, 1 A is a typical household current. The charge on an electron is -1.6×10^{-19} C, so a current of 1 A corresponds to $1/1.6 \times 10^{-19} = 6.25 \times 10^{18}$ electrons per second. So, the first constant μ_0, called the permeability of free space, which sets the scale of the magnetic fields in the system of units we use, is exactly (because we defined it that way)

$$\mu_0 = 4\pi \cdot 10^{-7} \cdot \text{N A}^{-2} = 4\pi \cdot 10^{-7} \cdot \text{N} \cdot \text{s}^2 \text{ C}^{-2}.$$

We now know what a coulomb is. If we go to a laboratory and measure the force F between two electric charges, Q_1 and Q_2, separated by some distance r, we find that $F \propto Q_1 Q_2/r^2$. Now, to make this an equation we need to measure the proportionality constant because we know how charge and length are measured. Doing this, we find that $F = Q_1 Q_2/(4\pi\varepsilon_0 r^2)$; this is called Coulomb's law. Note that we have chosen to write the proportionality constant (which we have measured) as $1/(4\pi\varepsilon_0)$. So, the second constant ε_0, called the permittivity of free space, which sets the scale of electric fields in the system of units we use, is exactly (because we measured it)

$$\varepsilon_0 = 8.85 \times 10^{-12} C^2/(N \cdot m^2).$$

Maxwell's equations predict waves which have a velocity of $1/\sqrt{(\varepsilon_0\mu_0)} = 3 \times 10^8 \text{ m s}^{-1}$. This is truly one of mankind's most remarkable intellectual achievements!

Physics is ...

The *Physicist* explores attributes of physics

Appendix B

A simplified derivation of the speed of light from Maxwell's Equations

Maxwell's equations in empty space:

$$\nabla \times \boldsymbol{E} = -\frac{\partial \boldsymbol{B}}{\partial t}$$

$$\nabla \times \boldsymbol{B} = \epsilon_0\mu_0\frac{\partial \boldsymbol{E}}{\partial t}$$

$$\nabla \cdot \boldsymbol{B} = 0$$

$$\nabla \cdot \boldsymbol{E} = 0.$$

Now, take the curl of the first equation

$$\nabla \times (\nabla \times \boldsymbol{E}) = -\frac{\partial(\nabla \times \boldsymbol{B})}{\partial t} = -\epsilon_0\mu_0\frac{\partial^2 \boldsymbol{E}}{\partial t^2} = -\frac{1}{v^2}\frac{\partial^2 \boldsymbol{E}}{\partial t^2}.$$

This is the three-dimensional wave equation with $v = \frac{1}{\sqrt{\epsilon_0\mu_0}}$.

It is easier to see if the electric field has only an x-component, $\boldsymbol{E} = 1_x E_x$. Then, calculating the curls of $\boldsymbol{E}$, you find that

$$\frac{\partial^2 E_x}{\partial x^2} = -\epsilon_0\mu_0\frac{\partial^2 E_x}{\partial t^2} = -\frac{1}{v^2}\frac{\partial^2 E_x}{\partial t^2}.$$

Therefore,

$$v = \frac{1}{\sqrt{\epsilon_0\mu_0}}.$$

It is pretty easy to show that *any* function $f(x - vt)$ satisfies the wave equation for E_x. For example, $f(x - vt) = A\sin(kx - \omega t) = A\sin(k(x - (\omega/k)t) = A\sin(k(x - vt))$ is a sinusoidal traveling wave.

doi:10.1088/978-1-6817-4445-2ch8

Appendix C

Galilean and Lorentz transformations

If one inertial frame (x, y, z, t) is at rest and another (x', y', z', t') moves in the $+x$ direction with speed v, and at $t = t' = 0$ the origins are coincident, the equations of Galilean relativity are:

$$x' = x - vt$$
$$y' = y$$
$$z' = z$$
$$t' = t$$

Figure C.1 shows a blue ball with its x-coordinates indicated at time $t = t'$. The ball also has velocity u and acceleration a, both in the x-direction. To find u' and a' is simply a matter of applying the definitions of velocity and acceleration:

$$u' = \mathrm{d}x'/\mathrm{d}t' = \mathrm{d}/\mathrm{d}t(x - vt) = \mathrm{d}x/\mathrm{d}t - v = u - v$$

$$a' = \mathrm{d}u'/\mathrm{d}t' = \mathrm{d}u'/\mathrm{d}t = \mathrm{d}u/\mathrm{d}t - \mathrm{d}v/\mathrm{d}t = a - 0 = a.$$

Note that the assumption that is most wrong is that clocks in both coordinate systems run at the same rate.

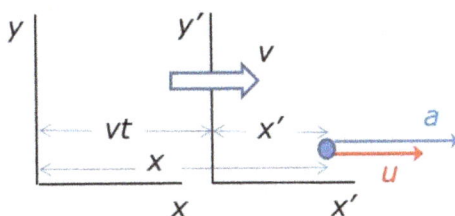

Figure C.1. Galilean transformation.

C-1   © Morgan & Claypool Publishers 2016

The corresponding Lorentz transformations are:

$$x' = \gamma(x - vt)$$
$$y' = y$$
$$z' = z$$
$$t' = \gamma(t - vx/c^2)$$
$$u' = (u - v)/(1 + uv/c^2)$$

a' is too complicated to write and of no particular use.

Appendix D

Gravitational field of a cylinder

Figure D.1 shows how, for cylinders that are longer than they are thick, forces from symmetrically spaced pairs of masses mostly cancel out on the 'equator'.

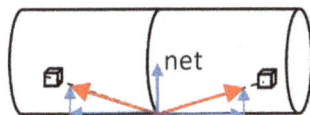

Figure D.1. Cancellation of axial components at the equator.

The on-axis gravitational field g of a thin disk of mass M and radius R and a distance x from the center of one end is $g = (2GM/R^2)[1 - (x/\sqrt{(x^2+R^2)})]$. The cylinder may be viewed as a stack of infinitesimal disks, as shown in figure D.2.

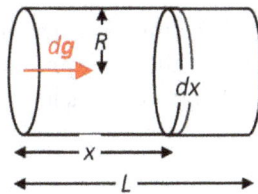

Figure D.2. Finding the fields at the poles.

The mass density of the cylinder (assuming it is uniform) is $\rho = M/(\pi R^2 L)$, so the mass of the disk of thickness dx is $dM = \rho \pi R^2 dx = M dx/L$. So, the field dg due to the thin disk is

$$dg = \left(2G dM / R^2\right)\left[1 - \left(x/\sqrt{(x^2 + R^2)}\right)\right] = \left(2GM/(LR^2)\right)\left[1 - \left(x/\sqrt{(x^2 + R^2)}\right)\right]dx.$$

doi:10.1088/978-1-6817-4445-2ch10

D-1

To get the net field, just integrate from $x = 0$ to $x = L$. The result is

$$g_{\text{end}} = \left[GM/(RL) \right]\left[(2L/R) + \tfrac{1}{2} - \tfrac{1}{2}\sqrt{(R^2 + L^2)}\big/R \right].$$

For
$$L \gg R, \; g_{\text{end}} \approx \left[GM/(RL) \right]\left[(2L/R) - (\tfrac{1}{2}L/R) \right] = 3GM/(2R^2).$$

Gauss's law for gravitational fields is $\int \mathbf{g} \cdot \mathrm{d}A = 4\pi GM_{\text{enc}}$, where M_{enc} is the mass enclosed by the Gaussian surface. Refer to figure D.3.

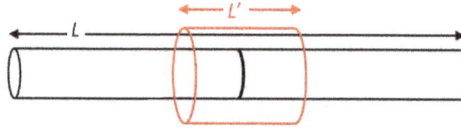

Figure D.3. Gauss's law applied to a long thin cylinder.

If the length of the cylinder is much longer than its radius, the field near the center will be very nearly constant and radial. So an appropriate Gauss's surface would be a concentric cylinder of radius r and length L', where $L \gg L'$. As shown above, the mass density is $\rho = M/(\pi R^2 L)$, and so $M_{\text{enc}} = ML'/L$. So,

$$\int \mathbf{g} \cdot \mathrm{d}A = 4\pi GM_{\text{enc}} \approx g(2\pi r L') = 4\pi GML'/L.$$

To get the field at the surface, let $r = R$:

$$g_{\text{equator}} \approx 2MG/(RL).$$

Now compare g_{end} with g_{equator} for $L \gg R$:

$$3GM/(2R^2) \gg 2MG/(RL).$$

Therefore the field at the ends will be much larger than the field at the equator.

If $R \gg L$, the field at the center of the end approaches the field in the center of a disk which is zero by symmetry. Therefore the field at the edge will be larger than the field at the ends.

Appendix E

Newtonian mechanics

E.1. Newton's first law

Imagine a book sitting on a table. It is at rest. There are two forces on it, its own weight (the force of the Earth pulling it down) and the force of the table keeping it from falling to the ground. Newton's first law simply states that, because the book is at rest, the magnitude of the weight (pointing down) must be equal to the magnitude of the table force (but pointing up), or, to put it more elegantly, the net force is zero. If you were pushing with a force that was just right so that the book moved with constant speed across the table, there would be two new forces on the book, the pusher pushing and the table resisting (called friction); but all the forces (now four of them) on the book still add up to zero. An object which is at rest or moving with constant speed in a straight line is said to be in equilibrium. Newton's first law can be expressed as follows. *The net force on an object in equilibrium is zero.* This law may seem obvious today, but it was revolutionary when Newton first stated it. Before Newton, it was assumed that the natural state of an object was to be at rest and, in order to keep something moving, there had to be a force pulling or pushing it.

When we get to Newton's second law, the first law will seem to be an unnecessary special case of the second. But the first law plays a much more important role than that. Suppose we ask the question, 'Is a law of physics always true?' The answer, perhaps surprisingly, is no; there are usually conditions under which a law is true. Thinking about the first law, imagine that you are inside an accelerating car. You are at rest *inside* the automobile (in equilibrium according to the first law), but you feel the seat back pushing forward on you, an unbalanced force. Therefore Newton's first law is untrue in this automobile. Whenever you find that Newton's first law is true you are in what is called an *inertial frame of reference*. This is the more important role played by the first law, as a test of whether Newton's laws are true laws for you.

When you have found one inertial frame, you have found them all, because any frame which moves with constant velocity relative to another is also an inertial frame. Any frame which accelerates relative to an inertial frame is not one. Because the Earth rotates and revolves around the Sun, it is accelerating and not an inertial frame. Fortunately, the accelerations involved are small enough to be negligible for many examples in everyday life.

Incidentally, inertial frames get their name from an alternative name for the first law, the *law of inertia*. Inertia means unwillingness to change and the first law says that you need to push or pull on something at rest or moving with constant velocity to change its motion.

E.2. Newton's second law

In order to discuss the second law, a brief detour to discuss units is imperative. Whole books have been written on this topic and I will be brief, assuming the reader already has a good sense of how we normally measure length, mass, and time. I will usually use SI units, mass is a kilogram (kg), length is a meter (m), and time is a second (s). These are all operationally defined in rather complicated ways, but it is sufficient to simply think of a meter stick and a stopwatch for almost everything which will be discussed in this book. 1 kg = 1000 g and 1 g is the mass of 1 cm^3 of water; or, you might like to think of a kilogram as having a weight of about 2.2 lb if you are in a country which uses British units. Note that I have not talked about how force is measured; that is part of what the second law is all about.

A brief discussion of acceleration is also in order. Everyone is comfortable with what velocity is, the distance traveled divided by the time to travel it; a scientist would call this the rate of change of position. Almost nobody, in my experience, is really comfortable with what acceleration is. It is simply the rate of change of velocity. A dropped ball, for example, gains about 10 m s^{-1} in velocity for each second it falls; after 1 s it is falling with a speed of 10 m s^{-1}, after 2 s with a speed of 20 m s^{-1}, after 3 s with a speed of 30 m s^{-1}, etc. The acceleration is $a = 10$ (m s^{-1}) s$^{-1} = 10$ m s^{-2}.

To discover Newton's second law you must interact with nature. We will never find physical laws by just sitting at a desk and thinking; we must make measurements which tell us how things happen. Newton's second law is about how exerting a force on a mass changes its motion, i.e. how force, mass, and acceleration are related. The experiment I propose is pretty simple: push or pull on a mass with a force and measure the acceleration. First vary the mass and hold the force constant, then vary the force and hold the mass constant. Hopefully, the resulting data will lead to some general law. But there is a problem. At this stage, force is a qualitative concept, a push or a pull, and if we cannot measure it then how can we vary it or even hold it constant? But, I can imagine having a machine which always exerts the same force. I could push with my hand using the muscles in my arm with very roughly the same force each time. Or, I could attach a spring to the mass and always pull with a force such that the spring was always stretched by the same amount, an improvement over my hand/arm machine. Oh, and I will call the force my machine exerts 1 Baker (B). So, let's do the first part of the experiment with one of my

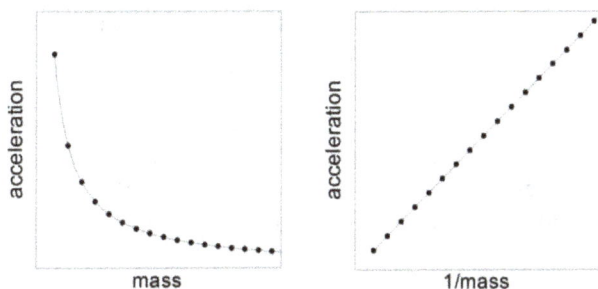

Figure E.1. Typical data, in arbitrary units.

constant-force machines, pulling with a force of 1 B, on various masses. The data might look like the graph shown in figure E.1 (left). Note that this makes sense because if the mass is large, then the acceleration is small, and *vice versa*. The problem is that it is difficult to quantify the relationship between the two variables because the data lie on a curve, not a straight line. Suppose that, instead, we plot the acceleration as a function of the reciprocal of the mass (1/mass), as shown in figure E.1 (right). This gives us a straight line which, means the acceleration is proportional to the reciprocal of the mass, $a \propto 1/m$. Next, we hold the mass constant while we vary the force, first using one 1 B machine, then two, then three, etc. We would find that two give twice the acceleration that one does, three give triple the acceleration, etc. In other words the acceleration is proportional to the force, $a \propto F$. Simple algebra says that therefore $a \propto F/m$. To me, this is Newton's second law, a statement of experimental facts—acceleration is proportional to force and inversely proportional to mass. Since it is much more convenient to convert this to an equation, we introduce a proportionality constant C, $a = CF/m$. The choice of C determines how we will measure force. The most clever choice, of course, would be $C = 1$, resulting in $F = ma$, the way we usually see the second law written. One unit of force, called a Newton (N), is that force which, when applied to an object with a mass of 1 kg, results in an acceleration of 1 m s^{-2}, $1 \text{ N} = 1 \text{ kg} \cdot \text{m s}^{-2}$ and is approximately 0.225 lb. In conclusion, the second law is both a statement of an experimental fact and a definition of a unit of force.

E.3. Newton's third law

The third law essentially says that forces in nature always appear in pairs and that *if object A exerts a force on object B, object B exerts an equal and opposite force on object A*. I will refer to those two forces as a Newton's third law pair and they always add up to zero. If you think about it, you will see that an alternative way of stating the third law is that *the net force on an isolated system of interacting objects is zero*, where an isolated system is one which has no forces acting on it other than the forces among its members.

There is often great confusion surrounding the third law. Carefully note from my first statement of the third law that the forces of a third-law pair are never on the same object. One of the questions cited at the beginning of this chapter wondered how we could move a book across a table, since the action and reaction force always cancel out. But only one of those forces is on the book and only forces on the book determine how the book moves. For the same reason, we should not make the mistake of identifying equal and opposite forces automatically as third-law pairs. For example, the weight of a book sitting on a table points down and the force the table exerts on the book is equal but points up; these are equal and opposite because of the first law, they have nothing to do with the third law.

E.4. Linear momentum

Newton, in his landmark book *Philosophiæ Naturalis Principia Mathematica*, did not write the second law as *F=ma*, rather he said that the rate of change of motion is equal to the force. It will be important to understand what this means if we are to understand many of the Q&A examples in this book. Recall that the acceleration is the rate of change of velocity; it is customary in mathematics and physics to write this as $a = \Delta v / \Delta t$, where Δv is the change in velocity and Δt is the elapsed time. For example, if the velocity increases from 4 m s^{-1} to 8 m s^{-1} over a period of 2 s, the acceleration is $(8 - 4)/2 = 2$ m s^{-2}. So now we can write $F = ma = m(\Delta v / \Delta t) = \Delta p / \Delta t$, where $p = mv$ is what Newton meant by the 'motion'; p is called the *linear momentum* today. (Note that since m is constant in $F = ma$, $m\Delta v = \Delta(mv)$.) If the net force on a collection of objects is zero, the rate of change of linear momentum must be zero—linear momentum never changes! This is called *conservation of linear momentum*. Conservation principles are extremely useful in physics. Note that conservation of linear momentum implicitly invokes the third law, since the net force on an isolated system must be zero.

E.5. Energy

The mathematics behind the idea of energy is more complex and will be handled in appendix F for the interested reader. The idea is that if a force is exerted on an object as it moves through some distance, work is done on the object which changes its energy. For many situations, the force is constant and along the path of the object so that the work can be written simply as $W = Fs$, where W is the work and s is the distance traveled. Now, what changes if you do work on an object? Well, that is really an easy question to answer qualitatively if you understand the second law—if you push in the direction it is moving, it speeds up, and if you push opposite the direction it is moving, it slows down. In other words, force causes acceleration, which means either speeding up or slowing down in physics, so what changes is speed. Without any derivation (see appendix F), here is the way that speed changes: $W = \Delta(\frac{1}{2}mv^2) = \frac{1}{2}mv_{\text{final}}^2 - \frac{1}{2}mv_{\text{initial}}^2$. This is often called the work–energy theorem. When you do work, you change the quantity $K = \frac{1}{2}mv^2$, which is called the kinetic energy of the object. Note that if there is no work done on a system, its kinetic energy never changes. Again, we have discovered a conservation principle, conservation of energy, which states that *a system on which no forces do work will have its total energy constant*. Something called potential energy is useful, but for the most part it will not be needed for this book. I will briefly discuss and define potential energy in appendix F and write the potential energy for weight, mgy.

Finally, the unit to measure energy and work is the joule (J), $1 \text{ J} = 1 \text{ kg} \cdot \text{m}^2 \text{ s}^{-2}$. You are also probably familiar with the unit of power, the rate at which energy is used or created, the watt (W). $1 \text{ W} = 1 \text{ J s}^{-1}$. A 100 W light bulb consumes 100 J of energy each second.

Physics is ...

The *Physicist* explores attributes of physics

Appendix F

Energy

Although I use calculus so that I have generalized to forces which might vary, I do the calculations in one dimension for clarity. First, derive the work–energy theorem in Newtonian physics:

$$F = \mathrm{d}p/\mathrm{d}t$$
$$= m(\mathrm{d}v/\mathrm{d}t)$$
$$= m(\mathrm{d}v/\mathrm{d}x)(\mathrm{d}x/\mathrm{d}t)$$
$$= m(\mathrm{d}v/\mathrm{d}x)v.$$

Rearranging,

$$F\,\mathrm{d}x = mv\mathrm{d}v.$$

Integrating,

$$W = \int F\,\mathrm{d}x = m \int v\mathrm{d}v = \tfrac{1}{2}mv_2{}^2 - \tfrac{1}{2}mv_1{}^2 = \Delta K.$$

In the theory of special relativity everything is the same, except p is redefined:

$$F = \mathrm{d}p/\mathrm{d}t$$
$$= m\left(\mathrm{d}\left[v/\sqrt{(1 - v^2/c^2)}\right]\Big/\mathrm{d}t\right)$$
$$= m\left(\mathrm{d}\left[v/\sqrt{(1 - v^2/c^2)}\right]\Big/\mathrm{d}x\right)(\mathrm{d}x/\mathrm{d}t)$$
$$= mv\left(\mathrm{d}\left[v/\sqrt{(1 - v^2/c^2)}\right]\Big/\mathrm{d}x\right)$$
$$= mv\left(1 - v^2/c^2\right)^{-3/2}\mathrm{d}v/\mathrm{d}x.$$

Rearranging,

$$F\,\mathrm{d}x = mv\left(1 - v^2/c^2\right)^{-3/2}\mathrm{d}v.$$

doi:10.1088/978-1-6817-4445-2ch12

Integrating,

$$W = \int F\,dx = m \int v(1 - v^2/c^2)^{-3/2}\,dv = mc^2(\gamma_2 - \gamma_1) = \Delta K.$$

If the particle started from rest, $\gamma_1 = 1$, and ended at speed v, $\gamma_2 = \gamma = 1/\sqrt{(1 - v^2/c^2)}]$, then

$$K = mc^2(\gamma - 1).$$

Be sure to note that m is the rest mass. Now, this does not look much like $\frac{1}{2}mv^2$ for small v, so we need to look a little more closely. If $v \ll c$,

$$1/\sqrt{(1 - v^2/c^2)} \approx 1 + \tfrac{1}{2}v^2/c^2 + \cdots$$

This is just a binomial expansion, $(1 + z)^n \approx 1 + nz + \frac{1}{2}n(n - 1)z^2 + \cdots$ So now we can write

$$K \approx mc^2\left(1 + \tfrac{1}{2}v^2/c^2 + \cdots - 1\right) \approx \tfrac{1}{2}mv^2.$$

Physics is ...

The Physicist explores attributes of physics

Appendix G

Friction

When two surfaces are in contact with each other and sliding, they exert forces on each other. The force which is parallel to the surface of contact is called the frictional force. Friction can be exceedingly complicated, but for many real-world situations it is true that the frictional force f is approximately proportional to how hard the two surfaces are pressed together; that force is usually called the normal force N and is the force which the two surfaces exert on each other perpendicular to their surfaces. So, $f \propto N$. The simplest example is if the surfaces are horizontal so that $N = mg$ (because of Newton's first law). For any given situation you need to measure both f and N to find the proportionality constant to make this an equation, $f = \mu_k N$. μ_k is dimensionless (a ratio of forces) and called the coefficient of kinetic friction and, to a very good approximation for everyday situations, depends only on the materials in contact; μ_k is large for rubber on dry asphalt (0.5–0.8) and small for Teflon on Teflon (0.04), for example.

If the surfaces are not sliding, they may or may not exert forces parallel to the surfaces on each other. For example, if a book sits on a horizontal table, there is no frictional force. But, if you push gently horizontally, it does not move, so there must be a frictional force equal in magnitude but in the opposite direction to your force. As you push harder and harder, the friction gets bigger and bigger until, eventually, the book pops away. The maximum frictional force you can get, f_{max}, is again proportional to N and the proportionality constant is the coefficient of static friction, μ_s; $f_{max} = \mu_s N$. In general, the static frictional force may be written as $f \leqslant \mu_s N$. It is always true that $\mu_s > \mu_k$, because when f_{max} is reached the formerly static object accelerates.

Appendix H

Rotational physics

Newton's laws, as described in appendix E, applies to objects which can do nothing but move along lines in response to forces; this kind of motion is called translational motion. But, this is not the most general way objects can move. Imagine a stick which you have grasped at one end and thrown; the stick does indeed seem to follow a path like a small ball would, but it also spins like a propeller as it moves along that path. Objects can also rotate and Newton's laws need to be extended to include the possibility of rotation. The description of rotational physics often runs to two to three chapters in a physics textbook and most of the formalism is not needed here. It is, though, important for the purposes of this book to include some basic ideas.

Torque plays the role of force. Where a net force causes an acceleration (speeding up or slowing down), a net torque causes an angular acceleration (spinning faster or slower). Torque is a little trickier than force to understand because it matters where the force is applied and in what direction. My favorite way to explain it is to imagine closing a door. You must push on the door somehow. But suppose you push on the edge of the door where the hinges are—it will not close no matter how hard you push. Now suppose you push on the edge of the door opposite where the hinges are, but you push straight toward the hinges—it will not close no matter how hard you push. Torque τ is defined as the component of the force perpendicular $F_\perp$ to the moment arm L times the length of the moment arm, $\tau = F_\perp L = FL \sin\theta$. You must choose an axis around which to calculate torques. If you call the torque positive, as I have for the example in figure H.1, and it tends to cause the moment arm to rotate clockwise, as you can see it does, then every other torque which tends to cause clockwise rotation must also be positive, and torques which tend to cause

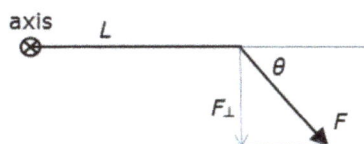

Figure H.1. A force exerting a torque.

counterclockwise rotation must be negative. Most of the questions answered are equilibrium problems, where all the torques add to zero. If there is a net torque, there will be an angular acceleration α and Newton's second law takes the form $\tau = I\alpha$, where I is the moment of inertia, which plays the role of mass in rotational physics. You need not worry about how to calculate I, it will always be given in any problems in this book. For example, the moment of inertia of a solid uniform disk of radius R and mass M is $I = \frac{1}{2}MR^2$. The name makes sense because if m is the resistance to acceleration (inertia), I is the resistance to angular acceleration (moment of inertia).

www.ingramcontent.com/pod-product-compliance
Lightning Source LLC
Chambersburg PA
CBHW082107210326
41599CB00033B/6614